BISON
BOOKS

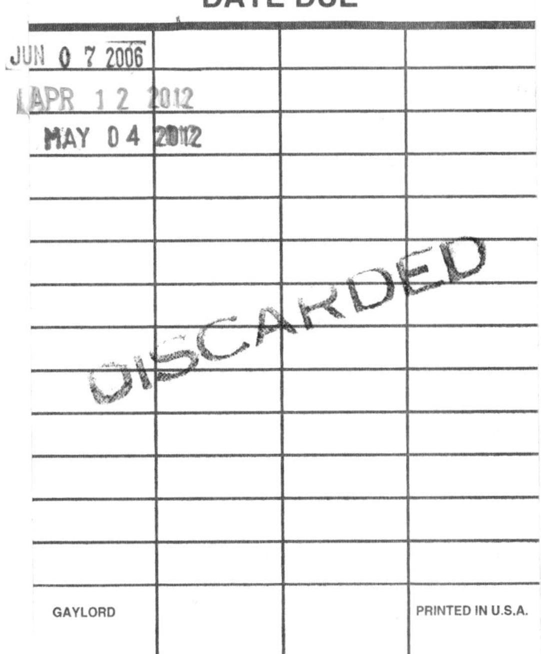
DATE DUE
JUN 0 7 2006
APR 1 2 2012
MAY 0 4 2012
DISCARDED
GAYLORD
PRINTED IN U.S.A.

VOLUME 17 IN THE SERIES

Our Sustainable Future

LESLIE A. DURAM

Good Growing

Why

Organic

Farming

Works

University of Nebraska Press, Lincoln and London

Manufactured in the United States of America. Publication of this book was made possible in part by a grant from the Organic Farming Research Foundation, Santa Cruz, California. Photograph on p. 117 courtesy of Glenda Kapsalis and photograph on p. 104 courtesy of Jon Bathgate; all others from author's personal collection. ♾
Library of Congress Cataloging-in-Publication Data.
Duram, Leslie A. Good growing: why organic farming works / Leslie A. Duram. p. cm. – (Our sustainable future; v. 17) Includes bibliographical references (p.) and index.
ISBN 0-8032-6648-0 (pbk. : alk. paper) – ISBN 0-8032-0496-5 (electronic) 1. Organic farming – United States. I. Title. II. Series. S605.5.D87 2005
631.5'84 – dc22 2004016267

For Jon, Kyle, and Maggie-ann

Contents

Illustrations

Preface

My interest in agriculture goes back to my childhood. I'm from Kansas. I love the open space of the plains, the endless sky, and the power of nature that is so obvious there. You hear a lot about farm foreclosures and smaller farmers "going under" when you live in an agricultural state. With these roots, I also grew up with a strong sense of needing to see the world. I became a geographer, which allows me to travel and study the interactions among the earth's people and environments. My initial academic work was on demographic studies in Europe, but I had this nagging feeling that my research didn't matter. I couldn't do anything to change things; I was an outsider. Finally, I realized that my true calling was back at my roots—plains agriculture.

Over the years, I have been inspired by several great geographers: Duane Nellis, Bill Riebsame Travis, Jim Wescoat, Steve White, and Gilbert White. They taught me to enjoy research, study what I truly want to study, and make it practical.

I began research on the plains of eastern Colorado (sort of an extension of western Kansas, really). As part of a project looking at conventional agriculture, I interviewed farmers. I loved it. I realized that the best way to understand something is to ask the people directly involved. But I also realized that something was terribly wrong. Conventional agriculture was going nowhere. I needed to understand how farmers could survive, and I discovered the answer was organic farming. It provides true opportunity for ecological and social sustainability within an agricultural system that is teetering on the brink of collapse.

I have researched organic farming for over a decade and been a conscientious eater far longer. Being both a mother and a researcher intensifies every page of agricultural information I read. What would be a mere agricultural statistic to someone else suddenly becomes a parental concern: What are my kids really eating? Does this food support family farmers who should be sustained into my children's futures? Will my kids have an opportunity to see a healthy rural environment?

Sadly, there is real need for concern. We should be worried about our food supply. Just as we live in an "industrialized" nation, we also eat industrialized global food. Fast food certainly, almost proudly, portrays itself as assembly-line fare – you can eat at any restaurant in the chain and get the same burger! Just read *Fast Food Nation* (Schlosser 2002) to see how fast food sickeningly controls American culture. But less obviously, and perhaps more dangerously, we Americans are eating food that is processed, distributed, and controlled by a global food industry. Farmers are far, far away from our dining tables, and they, too, are increasingly dependent on the agribusiness corporations that supply their inputs and buy the grains, vegetables, and livestock they produce. It's a far cry from the quaint little farms that we like to imagine dotting our countryside. Instead, there are huge industrial farming operations that produce a great deal of grain (most of which is fed to livestock), and intensively managed vegetable operations that demand the control of nature that pesticides offer, and dairy operations that inject their cattle with genetically engineered hormones to force them to produce more, and huge livestock containment operations housing thousands of animals that create tons of excrement. These industrial farms and corporate relationships can lead to ecological degradation (obviously, producing tons of manure per day is problematic), economic devastation (small and medium-sized acreages are often not profitable, and so many farmers go under), and social decay (rural communities are disappearing).

Is this the rural geography we want our children to inherit? Do we have a choice?

Indeed we, as individuals, can make better choices. We can try to avoid and alleviate the problems with industrial agriculture. One important choice that is gaining attention for various reasons is buying organic food. Consumers speak with their wallets, so first off we should buy what we believe in. And, clearly, buying organic food has the potential to address the ecological, economic, and social concerns that go hand in hand with our global industrial food system. So buying food that is produced locally and organically is the best way to "speak out" against these problems.

At the same time, the organic food system, as it is evolving in the United States, is wrought with potential problems. Is it really better to buy an organic apple produced on a large farm 2,000 miles away that has been distributed by a large corporation, or is it better to buy a locally grown apple from a farmer you know, even if it is not certified organic? No philosophical underpinnings are included in the standards by which organic products are certified, so an ecologically concerned small-acreage farm is not necessarily part of the mix. And the U.S. Department of Agriculture's National Organic Standards (effective 2002) are also solely related to production methods, not social concerns. Yet, I repeat, the best way to address problems with our global industrialized food system is to educate ourselves and buy the right food – to satisfy our personal goals of ecological and social sustainability. We must encourage and support family organic farmers who work outside the industrial production system. They have learned how to succeed within a system of agriculture that is set up for them to fail. This book is about organic farmers and their gutsy, innovative, successful efforts to beat the system.

This book would not have been possible without the help and generous participation of numerous organic farmers over the years. I've learned so much from you. Thank you all.

Thanks to my husband, Jon, whose calm has balanced my frenzy for so many years (although he is proudly a goofball in his own right). Hugs to my children, Kyle and Maggie-ann, whose smiles are a daily reminder of what is truly important in life. I thank the Southern Illinois University at Carbondale Department of Geography and Environmental Resources for allowing me to take the research directions that I choose.

OVERVIEW OF THIS BOOK

Geography is much more than maps; it allows us to investigate the complex social and ecological influences that create our landscape. Organic farming can play an increasingly important role in our rural areas. In chapter 1, key influences on organic farming in the United States are introduced: consumer demand, geographic approaches, social and ecological problems with industrial agriculture, the benefits of the organic farming alternative, research, information and certification on organic farms, and the current status of organic production in the United States.

Chapter 2 describes what scientific research has discovered regarding the many on-farm issues involved with organic farming: production comparisons, landscape qualities, and farmer decision making. In chapter 3, I present social issues through previous research on organic consumer behavior, rural food systems and communities, and the link between organic farming and social movements in agriculture.

Organic farmers can teach us a great deal about current agriculture. Chapter 4 looks at various types of operations in California, Colorado,

Florida, Illinois, and New York. Each farmer faces the on-farm production actions, the economic forces, and the social concerns of organic agriculture in a different way. In chapter 5, I discuss the key themes in organic farming, as discovered through interviews with these farmers. For readers who are not familiar with this system, I am providing citations within the text. So if you see (Jones 2000), that means that the source of my information is from an author named Jones who published in 2000. If you want to look up these specific documents, I have listed the full references at the end of the book.

What does the future hold? How will agriculture look in fifty years? Will future generations of family organic farmers have an opportunity to practice this method of production outside the industrial agricultural system? When organic production becomes "mainstream," will it change? Chapter 6 outlines the future of an alternative American agriculture – to be determined by consumer choices and farmers seeking independence. I make suggestions for promoting the comprehensive adoption of organic farming through new agricultural policies, more targeted research funding, educated consumers' actions, and informed advocacy.

Having indicated what this book does, let me also state what this book does not undertake. This is not a "how to farm" guide. While I provide examples of how some farmers have developed successful marketing systems, I do not outline a hands-on approach for all farms. Nor do I describe the specific in-field planting and pest-management techniques that lead to success. Rather, I describe successful organic cropping systems and explain how farmers gather information and make decisions.

Second, this book is a piece of advocacy scholarship, in which I use scientific evidence to support my assertion that our environment and society will benefit from a widespread conversion to true organic farming – carried out by family organic farmers. And I make suggestions for how this can be realistically achieved. So turn the page, settle in, and enjoy a geographical look at organic farming in the United States. I hope this book will educate you, stimulate your imagination, and convince you to seek alternatives to "the way things are." Educated consumers can make wise choices about the food they buy, and informed farmers can make the best choices for their land and their family. We all deserve something new: moderate-scale, regionally marketed family organic farms should be our goal. Why not make this mainstream agriculture?

1

Organic Farming and Geography

If we would divert to constructive research even a small fraction of the money spent each year on the development of ever more toxic sprays, we could find ways to use less dangerous materials and to keep poisons out of our waterways. When will the public become sufficiently aware of the facts to demand such action?
– Rachel Carson, *Silent Spring* (1962)

Geography is *geo* (earth) and *graphy* (to describe). This is a "geography" of organic farming because geography can best explain our complex world. Through words and graphics, geography can map the interrelated factors – both social (policy, culture, and economics) and ecological (climate, soil, water, and vegetation) – that influence our relationship to the earth. Specifically, a holistic approach is required in order to understand how organic farming is an integrated system of producing food and sustaining farm families. Furthermore, geography recognizes the many factors that make a *place*. And place matters. People are linked to the environment through their surroundings, at the local and regional levels and through our shared global environment. So our actions and decisions are related to place.

The geography of organic farming is evolving into a colorful patchwork of diverse farms across the United States. Regional variations can be seen: the eastern states, which generally tend to have more small-scale vegetable producers with local markets; the subtropical South, which has the appropriate climate for organic citrus production; the Midwest, which draws from its history of mixed livestock/grain operations; the plains, which have the space for organic grain production; and California, which has the experience and natural history to become a leader in marketing organic produce across the country. Although specific local conditions also come into play, these five geographical regions provide a framework for investigating organic agriculture. The multiple ecological and social issues that influence organic farming are played out in a unique way in each place.

This book presents organic farmers and their experiences within this regional geography. Successful organic farmers are undoubtedly the experts on organic farming. Pragmatism suggests that we view a situation through the eyes of those who know it best. So I let organic farmers speak for themselves. Through interviews and lengthy discussions they explain how they gather information, overcome adversity, find motivation, make management decisions, and take action. Their experiences and opinions provide a rich view of the topics relevant to organic agriculture today. This book also contributes a thorough survey of past research on organic farming. This research overview illustrates what topics are commonly studied and indicates what researchers and society as a whole think about organic farming. Much of this past research does not provide the rich geography of organic agriculture that organic farmers themselves reveal.

Overall, this book has four goals: to convince readers that a wholesale shift to organic farming would solve many of the problems that exist in U.S. agriculture; to offer extraordinary examples of innovative organic farmers who have successfully made this transition; to describe potential problems within organic agriculture (particularly Big O Ag: the large agribusiness corporations wanting to make a fast buck from the popularity of the organic label); and to outline clear actions that we must take to protect midsized family organic farms.

WHAT IS ORGANIC FARMING?

The term *organic farming* goes back to the 1940s when a British writer, Lord Northbourne, described an integrated farm as a "dynamic living organic whole" (Scofield 1986, 1). This idea of wholeness and complexity is still present within the definition of organic farms today (Høgh-Jensen 1998). Unfortunately, organic farming is often described as an opposite; it is defined by what it does not do (Tamm 2001). So organic farmers do *not* use synthetic fertilizers and pesticides and do *not* plant genetically engineered seeds. But what is the proactive definition? What *does* organic farming mean? According to organic agricultural researchers and the farmers interviewed for this book, it means crop rotation (changing the crops grown in a field each season) to build healthy fertile soil that has few pest problems (Watson et al. 2002). Organic farmers believe that "weeds are an index of the character of the soil," so spraying pesticides only treats "the effect, not the cause" (Walters and Fenzau 1996, xii). Organic farming means using "beneficials" – beneficial insects such as ladybugs that destroy the bad bugs like aphids, and beneficial interplanting of certain plants to keep pests away (Lampkin 1990). It means unique farm management decisions in terms of crop choice, planning, harvesting, and marketing (Gaskell et al. 2000). It means marketing through distinct channels – farmers must work hard to identify and maintain their sales outlets, often selling to numerous wholesalers, to brokers, or directly to consumers (Lampkin and Padel 1994). Marketing their farm products sometimes takes as much time as growing them, as organic farmers are trying to gain back the farmer's share of the customer's food dollar (decreasing from 40 percent in 1910 to only 10 percent today, according to Magdoff et al. 2000) by marketing directly to consumers. Organic farming also means diversity – growing a large number of crops both for ecological diversity and for sales diversity; not putting all your crops in one basket, so to speak (Newton 2002). It means independence – staying outside the mainstream industrial agricultural system as much as possible. And most certainly, it means innovation – trying new crop rotations or varieties or timing, trying new machinery (that they probably build themselves), and trying new sales venues to meet consumers' demands.

The term *certified organic* is important because it signifies a specific process of certification that has been regulated by the United States Department of Agriculture's National Organic Certification Standards since 2002. Accredited by the USDA, various state and regional certifying agencies (described later in this chapter) act to verify the field methods employed and to document the organic farming processes found on each farm. Farmers must forego synthetic agrichemicals for three consecutive years; they must maintain detailed farm histories; they must document every input to their fields; they must have an annual inspection by an outside inspector; and they must show that they are building their soil through rotation and use of green manure (crops planted and plowed under to fertilize the soil).

This book celebrates organic farmers and seeks to encourage broader acceptance of certified organic systems as part of a sustainable agricultural system. Organic farming, by going mainstream, could provide a unifying theme for the sustainability movement (*Nature* 2004), which in turn could help promote genuine organic methods based on family organic farms. There is strength in numbers, and together we could create a sustainable future through ecologically sound landscapes and viable rural communities; this begins with educated consumers.

ORGANIC GROWTH

By all accounts organic production and consumption is booming, not just in the United States but in many other countries as well. In Switzerland, 11 percent of farms are organic; in Austria this figure is 9 percent, and in Denmark it is 6 percent. Organic food makes up 4 percent of food sales in Denmark and Austria, 2 percent in Germany and Switzerland, and 1.5 percent in the United States (Organic Europe 2003; Thompson 2000). Organic production is actually being pushed by positive government regulation in many places. Several European Union countries subsidize farmers during their conversion to organic methods, assist in building organic marketing channels, and provide technical assistance and information specifically for organic farmers (Foster and Lampkin 1999; Padel et al. 1999). The USDA implemented National Organic Standards in 2002, which is the first federal regulatory attention given to organic production. In addition to this policy push, organic farming is being pulled by demand. Consumer demand is huge and increasing. Estimates place the growth of U.S. organic markets at 20 percent annually since 1990 (Natural Foods Merchandiser 2002). The year 2000 marked an interesting threshold for Americans: this was the first time that more organic foods were sold in mainstream supermarkets than in any other venue (with natural foods stores and direct marketing as runners-up). In fact, 72 percent of conventional grocery stores now carry some organic food (Dimitri and Greene 2002).

The leading organic foods sales are fresh produce, nondairy beverages, breads and grains, packaged foods, and dairy products. Amazingly, organic dairy items increased fivefold between 1994 and 1999 (Dimitri and Greene 2002), which is the result of consumers seeking to avoid rBGH, a genetically engineered hormone that is injected into cows to increase milk production (DuPuis 2000). Sales of organic snacks, candy, and frozen foods have increased by a notable 70 percent in recent years (Klonsky 2000). The increased consumer demand for all organic foods is likely linked to consumer concern about pesticide residues and genetically modified organisms (GMOs) in their food (Klonsky 2000; Kouba 2003). Pesticide residues on food come from on-farm pesticide use, postharvest pesticide use, pesticides on imported food, and banned pesticides that still persist in our environment (Kuchler et al. 1996). Eating certified organic food can help reduce uncertainties about our food supply (Leon and DeWaal 2002). A study by the Consumers Union shows significantly lower pesticide residues on organic compared with conventional food (Burros 2002; Goldberg 2002). Parental concern about the safety of their children's food has been an important motivation behind organic food purchases. A recent study shows that children who eat organic food have significantly lower levels of pesticides in their urine. This research indicates that "consumption of organic fruits, vegetables, and juice can reduce children's exposure levels from above to below the United States Environmental Protection Agency's current guidelines, thereby shifting exposures from a range of uncertain risk to a range of negligible risk" (Curl et al. 2003, 377).

In addition to organic foods providing a level of safety for consumers, there is evidence that organic foods have nutritional superiority. Comparisons of nutritional quality of foods are complex, as crop variety, cultural practices, and growing conditions vary from farm to farm and even among fields on a given farm. But a comprehensive survey of forty-one previous nutritional studies included 1,240 food comparisons that encompassed thirty-five vitamins and minerals (Worthington 2001). Although a wide variety of food crops were included in these comparisons, lettuce, carrots, spinach, and cabbage were common to most studies. This research shows that there are higher nutrient levels in organically grown crops compared with conventionally produced crops. Specifically, organic crops contain significantly more vitamin C, iron, magnesium, phosphorus, and useable protein (Worthington 1998, 2001). At the same time, organic crops had lower levels of nitrates and heavy metals than conventional crops. But these results should not be relegated to the laboratory; rather, we must consider how people's entire diets are affected. Research is just beginning to look at this very issue, by studying people's "excretion of flavonoids and markers of antioxidative defense." In other words, people who have eaten an organic diet seem to have more antioxidants in their system, as discovered through urine tests (Grinder-Pedersen et al. 2003, 5671). Given that most Americans eat conventionally produced food, the diminished vitamin and mineral content of this food could lead to long-term nutrition inferiority and adverse health effects.

The Soil Association is the leading organic certification agency and educational organization in Great Britain. Their inclusive report "Organic Farming, Food Quality, and Human Health" identifies clear benefits of organically grown food in terms of food safety and higher nutrient content (Heaton 2001). They found that organic foods have lower pesticide and nitrate residues and no increased risk of food poisoning compared with conventional food. Since organic livestock do not receive the massive amounts of antibiotics that conventional animals receive, they do not contribute to the overuse and potential resistance of microorganisms that conventional agriculture does. GMOs are prohibited in organic farming, which adds confidence to consumers seeking food safety. Primary and secondary nutrients are higher in organically grown crops (namely, vitamin C, minerals, and phytonutrients). This report concludes that consumers wanting to increase their vitamin and mineral intake, while simultaneously reducing their ingestion of pesticide residues, GMOs, nitrates, and artificial additives, should opt for organic food. Further, British consumers have been jolted by food safety scares, such as mad cow disease (BSE), which led to growth in organic production and consumption as a true alternative to these problems (Reed 2001). German researchers evaluated 150 comparative studies from 1929 to 1994 and found that organic foods provide a better option. Specifically, organic vegetables, particularly leafy, tuber, or root varieties, have much lower nitrate levels than their conventionally grown counterparts (Woese et al. 1997). This research also shows that livestock prefer organic feed. In an important Swedish review of the health benefits of organic food (Lundegårdh and Mårtensson 2003, 12), the authors conclude, "Organic foods can strengthen the immune system and other defense systems depending on an interaction between various favourable properties of organic foods. The balance between mineral nutrients, content of pesticides and other contaminants and the contents of secondary metabolites may be most important for beneficial effect."

Danish researchers show that organic food is higher in secondary metabolites (the nutrients one level below vitamins) and that this added nutrition would benefit human health more than nonorganic foods (Brandt and Mølgaard 2001). In a study comparing organic and conventionally grown strawberries, blackberries, and corn, Asami et al. (2003) found that organic food had higher levels of antioxidants like vitamins C and E (Byrum 2003). Flavonoids (measured in the study as total phenolic content) are well-known antioxidants, but they are much lower in conventional crops because plants naturally produce phenolics, chemicals that help them defend themselves from pests. In conventional food production, synthetic pesticides are applied, and this depresses a plant's natural defense mechanism. Flavonoids are reduced, as are the antioxidant properties of this food. On the other hand, organically grown plants still need their defense mechanism, so their levels of antioxidants are high. In fact, conventional strawberries, blackberries, and corn had 19 percent, 50 percent, and 58 percent fewer antioxidants, respectively, than their organically grown counterparts (Byrum 2003).

Because of nutritional superiority, food safety, fresher taste, and environmental concern, consumer demand is clearly strong and growing, as people are willing to pay more for organically grown food. Estimates of these price variations are commonly in the range of 10–30 percent over vegetables grown with pesticides (Sok and Glaser 2001). Consumers are demanding more organic foods, and there is an increasing acceptance of organic agriculture in the United States.

The 2002 U.S. Census of Agriculture asked questions, for the first time, about organic acreage. This is good news, as it points to a minimal acknowledgment of organic farming within our government. The questions are: "Of the total acres reported [above], how many acres were used to raise certified organically produced crops?" and "What was the value of certified organically produced commodities sold from 'THIS OPERATION' in 2002?" (USDA–NASS 2002, 3, 16). The fact that it has taken so many years to begin collecting data on this important segment of American agriculture is a sign of the USDA's overwhelming support of industrial agriculture, often at the expense of organic farming methods and foods. Until these data are published, we must continue to rely on estimates of organic acreage gathered through various state and private agricultural groups.

One government report draws from these diverse sources of information to estimate that 0.28 percent of total U.S. cropland is devoted to certified organic methods, but this amount doubled between 1992 and 1997 (Greene 2001) and continues to increase. In 2001, certified organic cropland totaled 2.34 million acres (Greene and Kremen 2003). There are wide variations by crop type, with approximately 2 percent of the major fruit and vegetable crops, apples, carrots, lettuce, and grapes, and 1 percent of all tomatoes, grown by certified organic methods (Greene 2001). For grains, these figures are much lower: only 0.1 percent of corn, soybeans, and wheat are organically grown. But substantial amounts of specialty grains are certified organic: spelt (37 percent) and buckwheat (30 percent) (Greene 2001). Geographic variation is seen among the states, as California, North Dakota, Minnesota, Wisconsin, Iowa, and Montana have the largest certified organic acreage (influenced by large areas of pasture and rangeland), and California, Washington, Wisconsin, Minnesota, Iowa, Pennsylvania, Ohio, New York, Vermont, and Maine have the largest numbers of certified organic farmers (Greene and Kremen 2003).

With these dual positive aspects of regulations pushing and consumers pulling farmers toward organic production, why have we only witnessed modest shifts toward organic farming in the United States? The answer is complex, involving historical and economic factors ingrained in U.S. society.

ORGANIC VERSUS CONVENTIONAL AGRICULTURE

While I do not want to oversimplify complex topics, I do want to be certain that we all understand how organic farming contrasts with conventional production. Some farmers who identify with the term *sustainable agriculture* may also employ some of the same methods as organic farmers. To put it concisely, organic farming methods are based on complex crop rotations that build soil health and employ only organic fertilizer (i.e., spreading composted manure, planting nitrogen-fixing legumes) and natural pest control (i.e., introducing beneficial insects, following useful crop rotation). These diverse crops also provide farmers with numerous crops to sell to different markets, often earning a price premium over conventional crops. Organic farms are operationally diverse, often mixing various types of crops and livestock so that the risk is spread out over more opportunities. In addition, organic farms build numerous distinct marketing channels, often relying on selling directly within the local community so as to keep more profit on-farm and to build integration of regional food production and consumption.

These activities contrast sharply with most American agriculture, which is often called "conventional" or "industrial." Conventional agriculture uses limited crop rotations and so must rely heavily on synthetic fertilizers (i.e., anhydrous ammonia), which kill much of the living biota present in the soil, and toxic chemicals, which kill most weeds and insects. These chemicals seep down into the ground and contaminate the groundwater; they also run off the surface of the fields during a rainstorm and pollute creeks, rivers, and even the ocean downstream. Some percentage of these chemicals remain on the food and can harm farm workers harvesting it and consumers eating it. Conventional farmers are caught in a vicious cycle in which they grow vast amounts of crops, and this overproduction leads to extremely low and falling prices. Then farmers need even higher yield per acre the next year, so they must use even more agrichemicals – and still many family farms go bankrupt. At the same time, the agribusinesses that sell farmers the chemicals and buy the low priced crops are becoming increasingly wealthy. Conventional farmers' incomes are so low that they must rely on government subsidies in order to continue to produce food. So why is the American taxpayer wholly supporting this illogical system of conventional food production, in which agribusiness is the only winner? Because people don't know the truth. The development of this illogical system of food production reflects our complicated relationship with the countryside.

Although we hold a romantic ideal of the rural way of life, most Americans have grown up in towns and cities; our only link to the food we eat is our weekly trip to the supermarket. People demand low priced food. This has worked quite well within the scheme of large-scale agricultural production – high yield and low cost seem to go hand in hand. But there are two problems with this system: it forces farmers onto an economic treadmill that decimates rural communities, and it causes serious environmental degradation.

RURAL DECLINE

Farmers trying to survive within the current U.S. agricultural system feel they must buy the latest machinery, plant the newest genetically engineered seeds, and apply the latest toxic agrichemicals in order to produce as many bushels per acre as possible. They can only sell these high yielding crops for low commodity prices. In the long run most family farms just barely break even. Between 1984 and 1998, the price that consumers paid for food increased just 3 percent, but the price that farmers received for their crops dropped an incredible 36 percent (Lauck 2000). Attempting to survive this economic crunch, many farmers are forced to take a disastrous jog on the "treadmill of production": the ever-increasing need for more land and higher yields, even though the economic and ecological sustainability of this system is short-lived (Cochrane 1993). Wendell Berry eloquently notes that once a farmer shifts to industrial agriculture, "the economy of money has infiltrated and subverted the economies of nature, energy, and the human spirit" (Berry 1977, 46). Indeed, conventional farming is far from nature and closely aligned to agribusiness and their latest technologies.

Questioning Agricultural Biotechnology

GMOs or genetically engineered (GE) crops are grown from seeds that have been genetically altered to have specific traits: for example, soybeans containing a gene that makes them immune to a specific herbicide or corn that has a toxic pesticide within it. The first GE crops were planted in the United States in 1996, and these now include biopharmaceuticals and "medical food" such as rice that has been engineered to contain human proteins (Cummings 2004.) Due to strong opposition in much of the rest of the world, particularly in the UK, Western Europe, and Japan, the United States plants 72 percent of global GE crops today (USDA–ERS–GMO 2003). Several African nations, including Zambia, have turned away American GE commodities given as aid, due to their uncertain environmental and health effects (*New York Times*, October 30, 2002).

The main GE crops are corn, soybeans, cotton, canola, potatoes, tomatoes, sweet peppers, peanuts, and sunflower (USFDA–GMO 2002). Forty percent of corn, 73 percent of cotton, and 81 percent of soybeans grown in the United States are GE (USDA–NASS–GMO 2003). Because most processed foods contain soy or corn (in their various forms such as soy lecithin or corn syrup), "many processed foods on U.S. supermarket shelves contain biotech ingredients" (USDA–Amber Waves 2003). In fact, 60–70 percent of nonorganic food in the supermarket contains GMOs (Cornell University 2003). Many dairy products have genetically altered ingredients: 70 percent of cheese in the United States is made with a GE enzyme, and a GE version of bovine growth hormone (rBST or rBGH) is commonly given to dairy cattle, so the milk we drink may also be altered (Cornell University 2001). Since 1999, the use of rBST has been banned in Europe (European Union 1999). Some GE crops are actually registered as a pesticide with the EPA, since they have been engineered to contain the pesticide (Pollan 2001a).

Currently, GMO foods are not labeled, despite the fact that 94 percent of Americans believe they should be (Hallman et al. 2003). In an ABC News telephone poll of 1,024 Americans conducted in 2001, 52 percent said they would be less likely to buy such food (ABC News 2003). Yet these questionable GMOs are in most of our processed food today. And consumers have no way of knowing whether they are eating gene-altered corn flakes, potato chips, peanut butter, or any other processed food, for that matter. Certified organic food is not grown from GMO seed, so buying organic food is the only way to attempt to avoid GMOs.

The steps for genetically engineering an organism involve the use of an "insertion package" that contains the desired new trait, a bacterium or virus to overcome the host gene's defenses, and a marker gene that is antibiotic resistant so that the insertion can be verified. Once this package is forced into the host's cells, an antibiotic is administered and the surviving cells should be antibiotic resistant and have the new trait. This is clearly different from traditional hybrid plants in which related species have been crossbred. GMOs cross the species barrier by putting fish genes into tomatoes or nut genes into soybeans, and they rely on viruses and antibiotic resistant bacteria to do so (Grogan and Long 2000).

Vast sums of money have gone into the research to develop some of these gene altered crops by a handful of multinational corporations: Monsanto, DuPont, Novartis, AstraZeneca, and Aventis (now Bayer Crop Science). These corporations obviously have a vested interest in promoting the acceptance, sale, and use of GE crops in order to realize the money they spent on their development. Thus they are spending millions of dollars on "feel good" campaigns to try to convince people that GMOs are beneficial (Jaffe 2001). But there are several current and potential problems with GE crops and their regulation. The FDA claims that there are no "special labeling requirements for bioengineered foods as a class of foods" because they have "no basis for concluding that bioengineered foods differ from other foods in any meaningful or uniform way" (USFDA 2001). The process for approval of a GE crop is the following: the corporation that developed the gene altered seed voluntarily supplies safety data to the FDA for review, then the FDA issues the statement that it has "no further questions . . . at this time" (Jaffe 2001). There are no long-term health or environmental studies on GE seeds, crops, or food. We are relying solely on the biotech industry's voluntary submission of its own testing and safety data.

Potential and realized problems with GE crops include unwilling spread of GMOs through cross-pollination; increased pest resistance and evolution of "superbugs" or "super weeds"; occurrence of food allergies due to unknown genetic materials in our food; varying nutritional content of GE foods; lack of freedom for farmers as they must sign proprietary agreements with the GE corporations; and reduction in export markets (Pollack 2003; *New York Times*, February 19, 2003; Jaffe 2001; Cummins and Lilliston 2000; Grogan and Long 2000; Pimentel et al. 1989a). U.S. exports have been in jeopardy because other nations have refused to accept GE crops. The British Medical Association "calls for a moratorium on the further growing of commercial GM crops in the UK until more research has been carried out into the long-term health and environmental consequences" (British Medical Association 1999).

Thus American farmers are caught on the industrial treadmill once again, being told that GMOs will earn more profit, but then realizing that GE yields are highly variable and require high input costs (Altieri 2001). In addition, export markets refuse engineered crops due to consumer uprising. Farmers are also faced with the issue of pollen drift, so if a neighbor plants GMO corn, their corn could "become" GMO by harvest time, due to cross-pollination. Finally, farmers are caught within the legal grips of the agribusiness giants again, as they are forced to sign proprietary agreements with the GMO corporations, promising to use only their brand of pesticides and noting that they are not allowed to save seed from one year to be used the next (Phillipson 2001). Federal patent law prohibits anyone from harvesting more than one crop from their GE seeds, thus assuring the corporations of "return customers," as farmers must buy more seed each year, rather than saving their own seed (International Center for Technology Assessment 2001). Farmers in the United States and Canada are being sued by the billion-dollar

GMO corporate giants when GE crops are found growing in their fields with no signed contract (Canadian Broadcasting Corporation 2003; Beingessner 2003). Further, these corporations have developed so-called killer genes that essentially cause a seed to commit suicide, so farmers cannot save seeds and plant them the following year. This type of control is immoral, as corporate profits may jeopardize our global food supply (Rural Advancement Foundation International 2003).

The Pesticide Action Network notes, "While many potential human health and environmental impacts are associated with these crops, testing has been remarkably inadequate" (Pesticide Action Network of North America 2003). With no long-term safety studies, we've introduced these new genetically altered materials to our environment and into our bodies. We simply do not have the facts on GMOs, yet we are currently conducting a massive experiment on you, me, the rest of society, and our ecosystems. Organic agriculture and buying organic food are the only way to avoid being part of this global experiment, being driven by the profit motives of several agribusiness and pharmaceutical corporations. A consultant for the biotech industry summed up the situation with an eerie premonition: "The hope of the industry is that over time the market is so flooded [with GMOs] that there's nothing you can do about it, you just sort of surrender" (Cummins 2001). The broader question is whether farmers will have to admit defeat to the powerful agribusiness interests that control much of industrial agriculture; this is determined by farmers' economic status.

Addicted to Government Subsidies

The treadmill of production and reliance on technology also explain farmers' heavy reliance on farm subsidies. The U.S. Farm Bill authorizes marketing assistance loans and loan deficiency payments (LDPS) that are available to farmers who grow eligible commodities: wheat, corn, grain sorghum, barley, oats, soybeans, minor oilseeds, rice, and cotton (USDA–FSA 1998). LDPS allow farmers to sell their eligible commodities for the loan rate if it is higher than the county-posted price for an eligible commodity, thus providing minimum security for farmers. Due to extremely low commodity prices (corn is currently selling at about $2 per bushel, although it costs nearly $3 to grow it), farmers must depend on government subsidies, and these subsidies mean that corporations can purchase corn cheaply for use in their various products, such as corn syrup in soda pop and snacks, to produce ethanol, and to feed the millions of livestock that Americans consume (Pollan 2002).

When he signed the 2002 Farm Bill, President George W. Bush stated,

"The Farm Bill will strengthen the farm economy over the long term. It helps farmer independence, and preserves the farm way of life for generations. It helps America's farmers, and therefore it helps America" (USDA 2002). Certainly, this sounds good. Unfortunately, the statement is false. The reality is that the Farm Bill continues to subsidize farmers for increasing their crop yields; LDPs are figured as the quantity of a commodity produced multiplied by the payment rate. The more you grow, the higher the subsidy. This upholds the high yield mentality of the treadmill of production that forces farmers to get bigger just to stay in business. By promoting megafarms and forcing out small farmers, the rural economy is weakened and the rural ecosystems are compromised. This will have devastating effects on rural communities and landscapes, which hurts America.

Half of farm income comes from the federal government ($23 billion in 2001). Our tax dollars provide half the income earned by American farmers. Incredible as that may seem, it's infuriating to note that most of this (61 percent) goes to the wealthiest 10 percent of farms, while smaller farms are barely eking out a living or may actually realize net losses (Williams-Derry and Cook 2000). Our taxes support farmers so they can sell their crops cheaply to agribusiness corporations that sell the food back to me and you, reaping huge profits in the process, mostly at the expense of family farms. Why are we so bamboozled? Because the agribusiness corporations and their hundreds of lobbyists have paid millions of dollars to our congressmen to convince them that this agricultural system "works" – all in the name of saving family farmers!

Concentration and Inequality in Agriculture

The most comprehensive information on the issue of agribusiness consolidation and inequity in agriculture has been gathered by Bill Heffernan and Mary Hendrickson (both in the Department of Rural Sociology at the University of Missouri) for reports to the National Farmers Union. These are available online at www.nfu.org. Incredibly high market concentration exists for many agricultural commodities: 81 percent of beef packing, 81 percent of corn exports, 80 percent of soybean crushing, 65 percent of soybean exports, and 61 percent of flour milling are controlled by the top four firms in each category (Hendrickson and Heffernan 2002a). This type of concentration is referred to as horizontal concentration; that is, one level of the food system is under control, or a few companies control from here to the horizon – as far as one can see! The second type of concentration is vertical control – when one corporation is a major player in all the levels of food production, distribution, and consumption. This is also occurring, as we can see from the example of ConAgra, whose multiple subsidiaries produce livestock feed, feed-out and slaughter cattle, process pork and broilers, distribute agrichemicals and genetically engineered products, transport grain, process food, and sell food under many labels including Healthy Choice, Hunt's, and Peter Pan Peanut Butter (Grey 2000; Heffernan 1999, 2000; Hendrickson and Heffernan 2002b). In fact, with sales of $23.8 billion and profit of $1.6 billion in 1998, ConAgra is second only to Philip Morris in U.S. food processing companies and getting close to the international giant, Nestlé (Heffernan 1999).

This market concentration has explicit implications for the integrity of rural life and our ecosystems. The book *Corporate Reapers* (Krebs 1992), in sections appropriately titled "a rural bloodletting" and "efficiency and ruthlessness," explains that the demise of rural America is not the upshot of free market capitalism but the result of agribusiness price fixing and deliberate anticompetitive strategies. First, this concentration increases farmers' dependence on specific inputs at set prices. To buy seed, fertilizer, and agrichemicals from one company may not provide the best management options (industrial farmers often get advice from chemical dealers). Second, it hurts farmers' earnings. If they can only sell to a handful of companies, they may not be earning a fair price (and in fact corporations tend to buy out hundreds or thousands of grain elevators in a geographical region, so farmers really have only one place to sell their grain). Third, it means your food is controlled by corporate profit motives. Our food dollars are concentrated in the hands of a few megacompanies. Can we trust that they are giving us the most nutritious food possible? Fourth, this concentration makes the U.S. food supply vulnerable to the market forces of a very few corporations (Magdoff et al. 2000). Is there any competition in the system? This is an odd question for a capitalist economy, but a valid one now. Perhaps food is a unique product that deserves a specific policy in order to safeguard society (Hendrickson et al. 2001). Fifth, what about our rural communities? With fewer, larger farms, and inputs purchased from huge outside corporations, and crops sold and distributed to national and international firms, rural economies wither and rural areas decline. Finally, what does vertical and horizontal integration mean for the environment? The corporations that control farm inputs and prices, distribution, processing, and sales are not place-based. They are geographically disconnected. And this lack of concern for a specific place makes environmental degradation quite easy, as it is faceless and placeless. Rather than local, small-scale "mom & pop" stores, we now have detached multinational corporations determining the future of agriculture.

We Pay the External Costs

An economics textbook provides a useful definition. When price fails to register all the costs associated with the production of a good (so costs are external to the market and "accrue to parties other than the immediate buyer and seller"), these are called external costs (McConnell 1984, 71). Negative externalities are the real production costs that are foisted off on society. In agriculture, externalities have these characteristics: these costs are neglected, there is a time lag, damage often occurs to groups with no voice, the identity of the producer is not known, and the economic and policy solutions are poor (Pretty et al. 2000, 114). There are many examples of agricultural externalities: for example, farmers spray pesticides that contaminate groundwater, streams, and rivers that flow to the ocean and cause eutrophication (a dead zone) at the river's mouth. Massive agricultural use of fossil fuels (both directly in tractors and indirectly in petroleum-based agrichemicals) increases air pollution and produces greenhouse gases that exacerbate human-induced global climate change. Likewise, clearing land for agricultural cropping means either tearing down trees or draining wetlands, both of which decrease habitats for wildlife and affect air or water quality. In addition, thousands of people become ill from Salmonella and other bacteria in our food due to the industrial production techniques. But the farmer doesn't pay for these things, and agribusinesses certainly don't, and we consumers don't pay for it in monetary terms, since food prices remain low. So who pays?

First, we have to ask: what are the real costs? Pretty et al. (2000) provide an excellent overview of the total negative external costs of agriculture in the UK, which they estimate to be $3.7 billion for 1996. This figure is based on the costs of contamination of drinking water (from pesticide, fertilizer/nitrate, phosphorus, eroded soil, and bacteria from livestock manure), damage to wildlife and habitats, emissions of gases, soil erosion and loss of carbon, and food poisoning. And the authors readily admit that their study includes only those externalities with monetary value, so numerous other goods and values are excluded. (How much is a viable rural community worth?) Research conducted more than a decade ago already indicated that the indirect costs of pesticide use in the United States, including ecological degradation and human illness, were between $1 and $2 billion annually (Pimentel 1991). We can only imagine how high this figure is today.

A more abstract idea, put forth by ecological economists, seeks to describe the value to be placed on what the environment does for humans. Personally, I find it a bit disturbing that we have to put a monetary value on nature to protect it. How ridiculous. We all know there is more to life than money. But perhaps these economists think that the best way to argue from an environmentalist's viewpoint is to put the environment into economic terms. Well, researchers have estimated prices for what the environment does for humans; they call this *ecosystem services*. Thus a wetland acts to control flooding near a stream, which saves people's homes, and this may be worth $250,000 in a given region. The economists estimate and then add up all the various benefits (or services) accomplished by the environment. Figuring values for seventeen ecosystem functions for sixteen biomes, Costanza et al. (1997) estimate that for the entire earth, the estimated value of ecosystem services is about $33 trillion per year. As a reference point, the authors note that annual global gross domestic product (that is, all the money that exchanged hands for goods and services) was $18 trillion. Thus the environment provides far more valuable services than we provide ourselves.

Agroecology is an interrelated system that encompasses many ecological concepts within an agricultural context (Altieri 1987) and suggests that ecological principles be used to guide farm management (Gliessman 1998). Rather than divide each component into separate parts, the goal of agroecology is to look at a farm as an entire working system – a whole ecological unit that also acknowledges social, ethical, and economic influences as well (Francis et al. 2003). This is particularly appropriate for organic farming, as management goals include long-term crop rotation, diversity, and interactions among plants, soils, insects, worms, and other key members of the community. Energy, water, and nutrient processes all occur independently and in unison over time, and if diversity flourishes, succession occurs so that a farm is constantly evolving toward a more complex biological state (Altieri 1987). Organic farms seeking this diverse, interactive approach will reach a high level of complexity that should benefit the natural ecosystem and the surrounding rural landscape. Organic farms provide enhanced ecosystem services, although no economic value is currently awarded for these efforts (Cacek and Langner 1986).

Björklund et al. (1999) uses the agricultural landscape of Sweden as an example to show that as agricultural production has industrialized and become more specialized over the past forty years, ecosystem services have declined. Altieri (1999) notes that biodiversity is part of ecological services that can lead to self-sustaining soil fertility and crop productivity – if diversified, low-input farming methods are used. In other words, industrial agriculture reduces ecological processes that support rural landscapes, and society as a whole must pay for this environmental deterioration. But conventional agriculture is not only responsible for ecological damage; society is being harmed as well.

Destruction of Rural Regions

Rural areas are in decline. The numbers are alarming. As the entire U.S. population doubled over the past seventy years, the number of farmers dropped from 7 million to only 2 million; and as recently at the 1990s, we have been losing about 32,500 farms per year (Kimbrell 2002, 17). This means more than just a loss of rural culture and a way of life. This implies vast changes for the rural landscape and communities. As these farmers went bankrupt, many were forced to sell out to their larger, more profitable neighbors. Average farm size in Illinois, for example, went from 196 acres in 1959 to 372 acres in 1997. But even this figure doesn't tell the whole story, as many small "farms" are listed for tax purposes as agricultural land, when there is very little actual production taking place. A more telling example of industrial agriculture is this: In 1959 Illinois had 129,157 farms that were between 50 and 999 acres. By 1997 this number dropped to 37,837. At the same time, the number of megafarms that are 1,000 acres or more went from only 574 in 1959 to 6,737 in 1997 (USDA–NASS 1997).

These numbers shock us but don't tell the human side of the story, as real families, communities, and towns are decimated by this type of agricultural change. The local stores and farming equipment dealers leave, then school districts consolidate, and eventually a small town can vanish. Rural people have to drive farther to a store, a post office, or a doctor, the quality of life diminishes, and soon the farmer's son and daughter want to leave rather than remain in the desolate area.

American history shows us that farming and rural life were vital to our country's development. "Jefferson believed that the system of land tenure and distribution adopted would ultimately determine the character of the new society" (Udall 1963, 32). A vigorous countryside provided a net of equality for all Americans, an obvious departure from the landed aristocracy of Europe at the time. So "he favored small freehold landownerships which would cause class distinctions to disappear. Growing as Jefferson would have had it grow, this country would have been a rural nation thinly populated by small farmers" (Udall 1963, 32). Our founding fathers and mothers would undoubtedly be shocked and disappointed to see the current level of our rural decline. And yet we, as a society, have idly sat by while millions of farmers have lost their land and left the countryside.

Today most of the money we spend on food does not go to the farmer who grew it; in fact, only about 10 percent does. The rest is profits for the input and marketing segments of agriculture. This squeezing of farmers, paying for seed, fertilizer, machinery, and chemical inputs, and earning so little per bushel or per pound, while the transportation and marketing firms earn so much, makes one wonder about the value of food in American society. "In other words, it may be difficult for a society that does not respect its farmers to respect its food" (Halweil 2000, 18). Taking that one step farther, it is easy to conclude that Americans have placed very little value on rural regions. Organic farming provides an opportunity for farmers to maintain rural life and for consumers to care about food, and for both to consciously decide to grow/buy organic food that will help sustain rural America.

In addition to the numerous social problems caused by the industrial agricultural system, there are also major concerns with agrichemicals. Pesticide use has complex ecological, economic, social, political, and ethical implications (Pimentel and Lehman 1993). Adoption of organic farming allows farmers to stop using these dangerous chemicals, which have such devastating human health and environmental effects.

PESTICIDE CONCERNS

It is important to understand the magnitude of the pesticide issue. According to the EPA, there are more than 865 active ingredients registered as pesticides that are used to create the thousands of pesticide products currently available. About 350 pesticides are commonly used on the foods we eat, in our homes, and on our pets (USEPA 1999).

People seem strangely trustful of modern agrichemicals. Even knowledgeable folks will say, "DDT was dangerous, but it has been banned for decades and new pesticides are much safer." The problem is that DDT (and numerous other chemicals) may have been banned, but it accumulates in the fatty tissue of living creatures and degrades only slowly, so even today most Americans still have detectable levels of DDT in their bodies. And sure, we can assume the newer pesticides are "safer" because they don't accumulate in the soil or our bodies for as long as the older (organochlorines or chlorinated hydrocarbon) chemicals. But we are now bombarded with hundreds of chemicals every day, and we do not know the cumulative effect of so many artificial substances – acting individually or interacting as a jumbled concoction within our bodies. Certainly we, as other organisms, can adapt to environmental change over generations, but not in a few years. Unfortunately, we don't have that much time. Just imagine little kids. Their small bodies must fend off multiple chemical exposures each day: insect repellant on their skin, weed spray at the park, bug killer sprayed in their school . . . not to mention the stew of insecticides, herbicides, fungicides, and fumigants that have left residues on their food. The bottom line is that *we do not know* the long-term combined effects of these multiple, constant chemical exposures, but we do know that many individual agrichemicals are dangerous and cause health problems.

Pesticides and Health

There are two ways to think of pesticide illness: acute and chronic. Acute exposures occur within a few hours or a day. Acute toxicity is commonly defined as the concentration required to kill 50 percent of laboratory test animals through either skin contact or ingestion (Trautmann et al. 1998). Acute poisonings lead to obvious, sudden illness that can range from allergic reactions to severe sickness or death. For each of the main types of agrichemicals, there are various reactions to acute poisoning (Moore 2002; Reigart and Roberts 1999). Nerve poison pesticides are organophosphates and methyl carbamates that are widely used around the world; acute exposures to these cause headache, dizziness, or even convulsions, coma, and death. Organochlorine pesticides are still used throughout the world, although several older compounds (such as DDT) have been banned; these pesticides cause allergic reactions, nausea, and convulsions. Pyrethrins are widely used in agriculture and home gardens; acute exposures can cause nausea, dizziness, headache, vomiting, and diarrhea. Dipyridyl pesticides like paraquat are commonly used worldwide and cause many acute poisonings each year. Symptoms include organ failure, lung damage, and pain; they can even cause death. Chlorophenoxy herbicides include common weed-killers such as 2,4-D and the base ingredient in Agent Orange; they cause skin irritation, headache, nausea, fever, irregular heartbeat, and mental confusion when acute poisoning occurs.

Chronic health effects of pesticide use may not be seen in humans until years or decades after exposure (Reigart and Roberts 1999); that is why it is so difficult to pinpoint the causes of illnesses related to agrichemicals. The organochlorine insecticides and chlorophenoxy herbicides are the main groups of chemicals associated with long-term health problems, such as cancer, neurological problems, developmental delays, reproductive disorders, and hormonal (endocrine) disruption (Moore 2002). Certainly, it is difficult to identify a precise pesticide exposure that may have occurred twenty or thirty years ago and prove that it caused the cancer present in someone today. But this complexity is also "convenient." The agrichemical corporations benefit from the time lag and the uncertainty of multiple exposures; these things make it very difficult to prove that a specific chemical is to blame. This is called the "benefits of chronic uncertainty" by some authors (Moore 2002). So while numerous scientific studies drag on, and government regulations seek to identify "safe" pesticide levels, the chemical companies are developing newer, "safer" pesticides to boost their profits. Meanwhile, the real issue of pesticide elimination is never addressed.

Here is an example of this "uncertainty" approach. Scientists from the Harvard School of Public Health conducted a study to estimate the dietary exposures to pesticides and heavy metals for 120,000 U.S. adults. They estimated people's annual diet by measuring frequency of eating certain foods (through a questionnaire) and potential contaminant residue data for table-ready foods (from FDA data) (MacIntosh et al. 1996). Although exposures to pesticides varied considerably by individual, "a substantial fraction of the population was estimated to have dietary intakes in excess of health-based standards established by the EPA" (MacIntosh et al. 1996, 202). But the authors caution that "before use for risk assessment or epidemiologic purposes, however, the validity of the exposure estimates must be evaluated by comparison with biological indicators of chronic exposure." And they concluded that "monitoring programs that use more sensitive study designs and population-based assessments for other subpopulations should be a priority for future research." These are gentle words to soften the harsh truth that we are all ingesting too many pesticides.

One aspect of the pesticide issue is undeniable: agrichemicals harm farmers' and farm workers' health. The incidence of cancer among farmers is greater than the population as a whole. One study shows that Iowa farmers are 25 percent more likely than people with other occupations to get leukemia or lymphoma (Rein 1992). Another study shows that farmers in Wisconsin have lower death rates related to tobacco and alcohol, but significantly higher rates of death from leukemia, lymphoma, and stomach, rectum, and eye cancers (Saftlas et al. 1987). "Modern chemical practices in farming" are linked to these increased rates, as "agricultural exposures were also positively associated with deaths" (119). Other studies show that farmers have an increased incidence of leukemia, Hodgkin's disease, non-Hodgkin's lymphoma, multiple myeloma, and cancers of the lip, skin, stomach, prostate, and brain (Novello 1991). One author notes that the "marked frequency of these cancers in farmers have not been conclusively identified," but "exposures to nitrates, pesticides, viruses, antigenic stimulants, and various fuels, oils, and solvents are suspected causes of many cancers" (Runyan 1993). Very few studies have investigated the effect of pesticide exposure to farm workers' health, because many workers are transient and rarely report health concerns to the local authorities. One recent study, however, used labor union and cancer registry data in California to find a strong correlation: exposure to certain pesticides increased farm workers' risk of prostate cancer 30–50 percent (Mills and Yang 2003). Such findings can only be viewed as conservative, given the level of underreporting among these workers.

Overall, agrichemical related illnesses are quite common among farmers, as reported by the large Agricultural Health Study of Iowa and North Carolina farmers. In fact, 7 percent of licensed restricted-use pesticide applicators have sought medical attention for chemical illness at some point (Alavanja et al. 1998). Farmers are mostly motivated to seek medical help because of acute poisoning, but the problem is also chronic, as blood samples from farm families show increased levels of pesticides that were used on the farm decades ago, but which have since been banned (Brock et al. 1998). Although we may not be aware of it, pesticide exposure occurs through many channels. For example, children of farm families are exposed to higher levels of pesticides than nonfarm kids, because these chemicals are found in their household dust (Simcox et al. 1995). In addition, studies show that birth defects and even infant deaths from these abnormalities are linked to pesticide exposures (Schreinemachers 2003).

Pesticide Regulation

The majority of U.S. agriculture follows the industrial, conventional system that relies on synthetic agrichemicals. Given the prominence of these pesticides and fertilizers, most people would assume that there is a central data bank that indicated the status of various pesticides in the United States. Well, this centralized, fundamental information does not exist. But with a great deal of poking and prodding, the various information sources can be found. The federal Food, Drug, and Cosmetic Act authorizes EPA to set maximum residue levels, or tolerances, for pesticides used in or on foods or animal feed. As far as human impacts, the act mandates primarily a health-based standard for setting the tolerance as "reasonable certainty of no harm." So the EPA follows a four-step process for human health risk assessment: Hazard Identification, Assessing Dose, Assessing Exposure, Risk Characterization (which is defined as Risk = Toxicity × Exposure). "Once EPA completes the risk assessment process for a pesticide, we use this information to determine if (when used according to label directions) there is a reasonable certainty that the pesticide will not harm a person's health" (USEPA 1999).

Most recently, the Food Quality Protection Act (FQPA) of 1996 required the EPA to begin considering new criteria before approving pesticides: considering exposures from all sources (food, water, residential), a methodology for trying to assess cumulative risk, and special sensitivity to children. In addition, FQPA requires the EPA to reassess tolerances for old pesticides by 2006, but in the meantime these chemicals are all still on the market, and we can only hope that the historical data on their safety is accurate (USEPA 1999). This gets to the heart of my information search – which pesticides have been banned in the United States, and whether they are still obvious in the environment today. According to the EPA, "Over time, registered pesticides, or certain uses of a registered pesticide, have been canceled. EPA does not maintain a listing of canceled pesticides" (USEPA–Canceled Uses 2003). Apparently, once a pesticide is banned, the EPA just pretends it never existed!

Partial information can be obtained through the EPA's international linkages. Luckily, the EPA is mandated by the Federal Insecticide, Fungicide, and Rodenticide Act to inform other governments about unregistered or banned pesticides exported from the United States that may affect importing countries (USEPA–UN PIC List 2003). In 1998, the UN Environment Programme and the Food and Agriculture Organization held the Convention in Rotterdam to address concerns about health and environmental risks associated with hazardous chemicals. Because only thirty countries have ratified the resulting document, it is still only voluntary. But the convention seeks to make a legally binding obligation that exporting nations list chemicals for Prior Informed Consent (PIC). According to the UN Convention, the export of a PIC chemical should take place only with the prior informed consent of the importing country (UN Report 2003). So the EPA lists sixty-four chemicals that are on the UN–PIC list; these are banned in the United States but still manufactured here and exported (USEPA–UN PIC List 2003). Looking at this list we see such pesticides as aldrin, chlordane, DDT, dieldrin, endrin, and heptachlor.

On the FDA Web site, there is a report called the *Pesticide Program: Residue Monitoring* 2000 (USFDA 2002). This presents the annual results of FDA sampling and testing of the U.S. food supply. Specifically, 6,523 samples of food were collected (2,525 samples of U.S.-produced food from forty-three states and 3,998 samples of food imported from eighty-two countries) and analyzed for pesticide residues. Residues were found in 40.4 percent of domestic samples and in 42.5 percent of the import samples. And of these detectable pesticides, samples exceeded violation levels in 0.7 percent of domestic samples and 3.8 percent of import samples, although these rates varied by food type (for example, 6.1 percent of imported vegetables violated tolerance levels). According to the report, "Many of the violative samples contained pesticide residues which were not registered in the United States for use in the commodities in which they were found; 15 domestic samples and 146 import samples fell into this category." Of the 396 pesticides that were detectable with the tests used, 117 pesticides were actually found in our food. This list includes banned pesticides such as BHC, chlordane, DDT, dieldrin, endrin, and heptachlor. Notice how this list of pesticides actually found in our food supply in 2000 matches the EPA's list of UN–PIC pesticides that are banned but produced in the United States for export. So we see that banning a pesticide for use in the United States does not mean that it will never be present in our food or our environment; quite the opposite. Chemical corporations simply sell these chemicals to developing countries that have lower environmental regulatory standards; the banned pesticides are used to produce food that is then imported by the United States. Weir and Schapiro (1981) call this the *Circle of Poison*. It is a prime example of corporations circumventing environmental laws for profit.

One Example: Aldrin

Do not assume that these banned chemicals are just harmless substances that really could safely still be on the market in the United States and elsewhere. Let's use aldrin as an example. Aldrin quickly breaks down to dieldrin in the body and in the environment. Both were widely used pesticides for crops like corn and cotton from the 1950s until 1974 when the EPA banned all uses except to control termites. Finally in 1987, EPA banned all uses according to the Agency for Toxic Substances and Disease Registry, which is part of the U.S. Department of Health and Human Services (ATSDR 2002). As mandated by Congress, ATSDR provides health information to prevent harmful exposures related to toxic substances.

Why did the EPA ban these chemicals? They determined that aldrin and dieldrin are probable human carcinogens. But that is not the end of the story. The various pesticide regulating agencies have studied these chemicals. The EPA set the tolerance for aldrin or dieldrin at a concentration of 0.0002 mg/L in drinking water; this level has been scientifically proven to limit a person's lifetime risk of developing cancer from exposure to each chemical to 1 in 10,000. The Occupational Safety and Health Administration (OSHA) has established the maximum average of 0.25 milligrams of aldrin and dieldrin per cubic meter of air (0.25 mg/m3) in the workplace during an eight-hour shift, forty-hour week. And the National Institute for Occupational

Safety and Health (NIOSH) also recommends a limit of 0.25 mg/m3 for both compounds for up to a ten-hour work day, forty-hour week. As noted above, the Food and Drug Administration regulates the residues of aldrin and dieldrin in raw foods, although the actual levels of aldrin and dieldrin in the samples were not noted. The FDA's allowable range is from 0 to 0.1 ppm, depending on the type of food product.

Aldrin is one of twelve persistent organic pollutants (POPs) that have been chosen as priority pollutants by the United Nations Environment Programme for their impact on human health and environment. It is one of the organochlorine insecticides that is persistent in ecosystems and accumulates in fatty tissues. "A growing body of scientific evidence associates human exposure to individual POPs with cancer, neurobehavioral impairment, immune system biochemical alterations and possible dysfunction, reproductive dysfunction, shortened period of lactation, and diabetes" (Orris et al. 2000, 7). POPs are also semivolatile, which means they can vaporize or be absorbed into the atmosphere. So these compounds can be transported great distances in the air and water (UNEP 2003). Some of the highest levels have been documented in both the northern and southern arctic areas (Orris et al. 2000). Thus, the impacts of industrial agriculture are geographically diffuse, reaching every location on earth.

"Inerts" Are Not

Not only are the active ingredients of our agrichemicals an issue of concern, but so too are the additional ingredients. The Federal Insecticide, Fungicide, and Rodenticide Act of 1947 established definitions that are still in effect today: an "active ingredient is one that prevents, destroys, repels or mitigates a pest, or is a plant regulator, defoliant, desiccant or nitrogen stabilizer." The active ingredient and its percentage weight must be listed on the pesticide label. On the other hand, "inert ingredients" are "any ingredient in the product that is not intended to affect a target pest." These inert ingredients are not listed on the label. In 1997, the EPA issued the Pesticide Regulation Notice 97–6, which encourages pesticide companies to use the words "other ingredients" because "it should not be assumed that all inert ingredients are non-toxic" (USEPA–Inert Ingredients 2003). In addition, many pesticides actually break down in the environment into very different chemicals that are just as toxic as their parent compounds (USGS 1999). For example, the commonly used herbicide atrazine breaks down into deethylatrazine (DEA), which may react very differently than the active ingredient itself.

Chemical corporations are only required to conduct safety tests of the "active" ingredients in pesticides. All the "inert" ingredients are never tested, with the excuse that this would divulge company trade secrets. The EPA does not require testing of actual pesticide products, but rather only the separate "active" ingredients. The EPA has identified four categories of these secret inert ingredients: toxicological concern; potentially toxic; unknown toxicity; and minimal risk/safe (USEPA–Inert Ingredients 2003). Thus we do not know the true health effects of off-the-shelf pesticide products as they are actually purchased and applied. Obviously, we don't want to conduct trials on humans, but a study of lab animals was alarming. Scientists tested common weed control products at extremely low levels and found a strong link to increased miscarriages (Cavieres et al. 2002). The amounts used in this study were lower than what the EPA mandates in product registration tests – parts per billion, which according to the authors is approximately one drop of pesticide in five hundred bathtubs of water. But even the lowest doses used had strong effects. Overall, it is clear that we need intensive research on these "inert" ingredients, which are present in all pesticide formulations. Our current knowledge base is miniscule at best, and this could lead to dangerous results.

Our Water

So far, I have focused on the direct human health-related aspects of pesticides from conventional agriculture. But the ecological aspects of pesticides and agricultural fertilizers are equally disturbing. These agrichemicals particularly impact our water supplies, both groundwater and surface water. Pollution is geographically widespread, as chemicals from any local farm flow into a nearby stream, which flows into a river, which flows into a reservoir, which provides drinking water for the region, and outlets into another river which, after thousands of miles, dumps into a bay and the ocean. In other words, one farmer could impact many waterways, many downstream states, and a large number of people. This exemplifies the complexity of non–point source pollution. Rather than one obvious effluent pipe spewing into a stream (a point source), agrichemical-laden water runs off many farm fields and flows into many streams and rivers.

Some of the best information on U.S. water quality is from the U.S. Geological Survey's National Water-Quality Assessment (NWQA) Program, a comprehensive study of fifty major river and aquifer systems that comprise watersheds covering half of the conterminous United States (USGS 1999). This research shows that "pesticides are widespread. At least one pesticide was detected in more than 95 percent of stream samples" and in over "60 percent of shallow wells sampled in agricultural areas" (USGS 2001, 6). But the occurrence of these chemicals is complicated by the fact that pesticides most often occur in mixtures, so "two-thirds of stream samples collected in agricultural areas contained 5 or more pesticides, and more than one-quarter of the samples contained 10 or more. Groundwater contained fewer pesticides; about 30 percent of the wells sampled contained 2 or more" (USGS 2001, 6). Another USGS study sampled groundwater in five states and found that 12 to 46 percent of wells in agricultural regions exceeded drinking water standards for nitrate concentrations, and also showed high levels of other agricultural inputs: potassium, chloride, calcium, and magnesium (Hamilton and Helsel 1995).

The seasonality of pesticide pollution is notable. Annual average concentrations may not exceed drinking water standards, but drastic violations of the standards may in fact occur for one or two months in the spring when pesticide application rates are high and rains increase the water run-off (USGS 1999). Overall, the study found that pesticides were found at low concentrations, mostly below drinking-water standards, but that the environmental and human risk from these chemicals is unclear. "For example, current standards and guidelines do not yet account for exposure to mixtures, and many pesticides and their breakdown products do not have standards or guidelines" (USGS 2001). Drinking water standards have been established for just forty-three of the seventy-six most commonly occurring pesticides studied by the Geological Survey (USGS 1999). Organochlorines pesticides (DDT, dieldrin, and chlordane) were all banned in the 1970s and 1980s, but they exceeded acceptable sediment limits in 20 percent of streams in agricultural areas (USGS 2001) and are still found in almost every fish sample today (USGS 1999).

Synthetic nitrogen fertilizers are commonly applied to conventional farmland, and nitrate levels are often higher in groundwater underlying farmland. "Nitrate concentrations in agricultural areas were among the highest measured" in the United States, and "nitrate in shallow groundwater was widespread and strongly related to agricultural land use" (USGS 1999). Specifically, "concentrations in about 20 percent of shallow wells sampled in agricultural areas exceeded the USEPA drinking water standard" (USGS 2001). High levels of nitrate in drinking water can cause "blue baby syndrome" in which oxygen levels in an infant's blood drop dangerously low (USGS 1999). This is of great concern in rural areas where wells are commonly used for household water supply; approximately 20 percent of the U.S. population relies on these wells (USEPA 1999). These domestic wells are not regulated, and many people do not know the quality of their well water. Information from Cooperative Extension Services and other offices seeks to educate rural residents about well water contamination from pesticides (Trautmann et al. 1998). They note that "between 1950 and 1980 the production of synthetic organic pesticides more than tripled in the United States" and "22 pesticides have been detected in U.S. wells." Another informational pamphlet notes that "the higher a pesticide's water solubility, the greater the amount of pesticide that can be carried in solution to groundwater" and "coarse, sandy soils with low organic content allow more rapid movement of surface water downward" (Dixon et al. 1992).

Establishment of drinking water standards is a complex process, with many estimates and assumptions. Data are often inadequate or complicated by the fact that laboratory animal data have to be used to estimate health effects in humans. Drinking water standards represent what authorities believe to be the acceptable level of risk to people exposed to chemicals in drinking water. The Safe Drinking Water Act of 1974 requires the EPA to set drinking water standards for public water systems that provide water to at least fifteen connections or twenty-five persons at least sixty days out of the year (USEPA 1999). Relevant to the concern about pesticides in our water, the EPA publishes data that list varying levels of cancer risks for chemicals found in drinking water. It is eerie information, reducing serious illnesses like cancer to abstract impersonal numbers. They also note contaminant levels for 10–4 Cancer Risk, which is defined as "the concentration of a chemical in drinking water corresponding to an excess estimated lifetime cancer risk of 1 in 10,000" (USEPA–Standards 2002). USEPA has set standards for approximately twenty-two agrichemical pesticide and nitrate contaminants (USEPA–Contaminants 2003), thus for most pesticides, drinking water standards have yet to be set.

The EPA has set a Maximum Contaminant Level Goal (MCLG) which is the contaminant level with no known health risk. But the enforceable standard is the Maximum Contaminant Level (MCL), which is the allowed amount of a contaminant, considering available technology and cost (USEPA 1999). According to the U.S. EPA's Safe Drinking Water Hotline, "Health advisories such as boil water notices or 'do not drink' orders are determined by state and local agencies. EPA does not have a national database of health advisory occurrences." The health advisory and violation data are collected at the state level, and most states submit reports to the USEPA (USEPA–Compliance 2000). Yet this does not provide an accurate view of agrichemical contamination for drinking water, because the results simply list reported violations of "synthetic organic chemicals" (USEPA – Factoids 2003) but not the specific pesticides, and of course many pesticides are not included because no drinking water standard has been set. One pilot research project was conducted jointly by the USGS and EPA using various laboratory methods to test water samples from twelve public water supplies for 178 different parent pesticide and breakdown products (Blomquist et al. 2001). They found 108 of these pesticide compounds in the sampled drinking water supplies.

Several issues of concern come to mind regarding our drinking water. First, there are no standards for mixtures of various pesticides (and the USGS water study shows that pesticides most often occur in groups). Second, there are many "variances" granted to water systems if they cannot afford to treat their water to reach lower contaminant levels (this is particularly relevant to pesticides, as it is very expensive to conduct the types of treatment necessary to remove them from the drinking water). Third, as noted in the USGS study, not all relevant pesticides are listed with standards (more pesticides must be under the EPA guidelines and "inert" ingredients should have standards). Fourth, the seasonal spikes in agrichemical use and pollution can overwhelm a water system (will they test for and catch the contaminant in time to issue a health advisory?). Finally, state collection of data is not centralized and does not necessarily follow through to USEPA guidelines (five states did not even file a report in 2000), so we must question the link between federal guidelines and state or local enforcement, particularly if "technology and cost" influence the regulation on whether treatment is mandatory for contaminant removal.

Indifferent or Misinformed?

All of this information on pesticides and agrichemicals is frightening, almost unreal. So why aren't Americans outraged by the dangers of pesticides and demanding clean, healthy organic farms and food? The best response I have is that we, as a society, do not support the "precautionary principle." The Europeans have a firm grasp on the precautionary principle, which is the basis of their opposition to genetically engineered food (Myhre and Traavik 2003). They feel that it is best to be cautious, that the safety of technology must be proven before they will accept its use. We Americans, on the other hand, have faith in science and technology. We trust that it will help us and be used wisely. We think, hey, it hasn't been proven to be bad, so let's just try it for a while, and it will probably work out just fine. In addition, media coverage on agrichemicals is nonexistent or slanted by corporate funding, so part of the problem is that we are not properly informed. We are not concerned because we don't have accurate information. Trusting technology is not necessarily bad, but the problem is that agricultural science and technology is under agricorporate control and their goal is profit. Hence safety is secondary.

This ingrained system is particularly baffling when we consider that at the same time the use of industrial agricultural pesticides soared (from 1965 to 1990) the estimated crop losses from insects, diseases, and weeds actually increased from about 35 percent to 42 percent worldwide (USDA–SARE 2000, 4). This indicates that while we are poisoning ourselves and our environment, we are also failing to control crop pests. Even before planting, seeds in industrial agriculture are treated with chemicals to halt seed-borne pathogens – despite the fact that recent research shows that simple hot-water treatments may be just as effective (Nega et al. 2003). Industrial agriculture reduces diversity by relying on just a few crops. This, in turn, simplifies and intensifies the pest complex present. Then pesticides kill off beneficial insects, making some pest densities higher and crop losses greater (Matson et al. 1997). In addition, pesticide resistance is an increasing problem. The most commonly used chemicals are being outsmarted by weeds, as they adapt and become immune to the once-effective sprays (Pollack 2003). In this evolutionary process, we can imagine the development of a "superweed" that survives and becomes the ancestor to future weed generations. This has serious long-term consequences for farmers worldwide.

The best way to step away from these risky synthetic pesticide and fertilizer hazards is to support organic farming methods. In addition to the obvious exclusion of synthetic pesticides that cause human health problems and water pollution in our environment, organic farming methods do not allow synthetic nitrogen fertilizers, which contaminate our water with nitrate and nitrite. Organic farmers use crop rotations, green manures (tilling under a crop to provide enriching compost), and livestock manure for fertilizing their soils. Scientists have found substantially lower levels of nitrate concentrations beneath organically cropped fields and high levels in soils under conventionally farmed fields (Honisch et al. 2002). Other research shows that organic farming could significantly reduce pesticide and nutrient run-off that pollutes regional and even international watersheds (Paulsen et al. 2002; Kersebaum et al. 2003).

Farmers and consumers alike see that organic farming is the best way to avoid the problems with pesticides and artificial fertilizers: water contamination, the question of "inert" ingredients, and the sheer magnitude and mixtures of dangerous chemical exposures today. In addition, farmers see organic methods as a means to distance themselves from the social problems of industrial agriculture: massive farm size needed to get higher yields, low crop prices, the game of federal subsidies, and the bankruptcies of many small to medium-sized family farms. A key component of the organic agricultural system is certification, as this provides the proof that organic farming methods are unique and produce distinct crops.

ORGANIC CERTIFICATION AND NATIONAL STANDARDS

Organic certification in the United States is carried out by state and independent agencies that act to collect paperwork, inspect, and grant certification to each organic farm and processor. In the past, each agency could have somewhat different standards (Fetter and Caswell 2002), so most farmers who sold their products nationally or internationally sought out the best known certifiers, as this provided them with increased credibility. Thus consumers in Kansas or Maine would still be accustomed to seeing CCOF (California Certified Organic Farmers – the oldest certifying agency) or Oregon Tilth certifications on their products – not that all ingredients were grown in Oregon or California, but rather that these certifications had gained respect and their labels were well known. But obviously, this was somewhat of a hodgepodge system of certification, with thirteen states and some forty-two private agencies active in organic certification in the United States (Duram 1998a). People realized that national standards should be established and recognized across the country and throughout the world.

So began the complicated relationship between the USDA and national organic standards. It has been a long process; the 1990 Farm Bill initiated the establishment of a USDA organic farming office to draft standards with the assistance of an advisory board (NOSB), but it wasn't until the mid-1990s that a draft document was published.

Secretary of Agriculture Dan Glickman walked a fine line – the USDA fears that endorsing organic production will signal problems with conventional agriculture that they are not willing to admit. So, on the one hand, Glickman noted, "It's time to take the next steps to fully embrace organic agriculture and give it a more prominent role in the farm policy of the 21st century" (Glickman 2000). But he also stated, "The USDA is not in the business of choosing sides, of stating preferences for one kind of food, one set of ingredients or one means of production over any other" (Kaufman 2000).

In addition to this complicated introduction, the proposed standards had many other problems. Although Glickman noted, "These are the strictest, most comprehensive organic standards in the world" (Kaufman 2000), others begged to differ. These draft standards caused an uproar in the organic and natural foods community – the proposed standards were too weak and seemed influenced by industrial agricultural interests. Specifically, the draft standards did not ban the use of sewage sludge, genetically modified materials, and irradiation in organic production. The USDA received nearly 300,000 public comments, mostly expressing concerns that the draft standards lacked safeguards and catered to a very watered-down definition of organic agriculture.

So the USDA went back to the drawing board and tightened the loopholes that were most obvious, and reintroduced the standards in January 2001 for public comment. These National Organic Certification Standards were eventually adopted and phased in over eighteen months. They went into effect on October 21, 2002. Farmers and consumers still have mixed feelings about them. They created another level of paperwork and bureaucracy for farmers' certifications, and they created some guidelines for consumers that are a bit confusing. There are now three designated levels of "organic-ness." Products labeled "100 percent organic" contain only organically produced ingredients. The "organic" label indicates products that are at least 95 percent certified organic. The designation "made with organic ingredients" is for items with at least 70 percent organic components, and up to three of these ingredients may be listed on the package. Any product containing less than 70 percent organic ingredients may not be marketed as an organic food. Foods that fall into the organic categories qualify to display the "USDA Organic" seal.

But despite the new logo, specific labels, and the new certification bureaucracy, the USDA presented the organic food information in such a low-key, negating manner, it's surprising that anyone heard about it. Even in the news release to mark the implementation of the new national standards, USDA Secretary Ann Veneman provided a carefully worded statement that now consumers will know that "products labeled as organic will be consistent across the country" (USDA News Release 2002) with no comment as to the benefits of organic production. The USDA pamphlet "Organic Food Standards and Labels: The Facts" specifically notes that "USDA makes no claims that organically produced food is safer or more nutritious than conventionally produced food. Organic food differs from conventionally produced food in the way it is grown, handled, and processed" (USDA–AMS 2002). Following this lukewarm introduction of the standards, a high-ranking USDA scientist reinforced this notion when she said that there is no evidence that one method of growing food is safer than the other (Reuters October 24, 2002). So, as always, it appears the USDA is carefully kowtowing to the large industrial agricultural interests in this country. Granted, they worked on the organic standards for twelve years, but they clearly seem to be saying that the standards don't really mean much.

The natural foods retail industry strongly supports the national standards. The Grocery Manufacturers of America represents food producers such as Kellogg's, Heinz, and Del Monte and they note, "standards bring a much needed uniformity"; whether consumers "live in California, Kansas, or Georgia," certified organic will have the same meaning (Pickrell 2002). Likewise, Whole Foods Market, the world's largest retailer of natural and organic foods, stated that the regulations come "at a time when shoppers are actively seeking out organic foods more than ever before" (Pickrell 2002). This highlights the conflict between the market-driven success of organic products and the grassroots ethical concerns of organic farming (Vos 2000). Further, the national standards will promote market stability and export markets, but will not necessarily support a locally based "socially and environmentally sustainable agriculture and food system" (DeLind 2000, 198). It is not clear how these divergent ideas will be balanced within the framework of the new regulations, but it seems that market growth is currently driving the process.

Well, for all this praise from the natural foods retailers, many organic farmers are still leery of the new standards. Some smaller organic growers feel that certification is not in their best interest, particularly if the costs for federal standards increases their fees. These farmers point out that the word *organic* is now owned by the USDA, and they need to search for a different term to describe their farming methods (*Growing for Market* 2002; Coleman 2002). On the other hand, for larger-scale farms the national regulations should provide uniformity that could strengthen consumer knowledge and demand for organic products. This may even increase perceptions of product reliability, and uniformity will expand export demand. Apparently, there still exists strong antiorganic (or at least organic neutral) feelings within the USDA, and the national standards do create more paperwork for each farmer. So the jury is still out on the national standards. The best case scenario is that we can say, after five years, that the standards have helped bolster demand and knowledge about organic products, and this has assisted some farmers in making a successful transition to organic farming.

RESEARCH AND INFORMATION

The National Standards represent an agricultural policy that obviously affects organic farmers, but they are also adversely affected by many indirect political actions. For example, many agricultural programs simply bypass organic farmers because their diverse operations don't qualify them for commodity assistance programs that provide key income for industrial farmers. Thus, U.S. government agricultural payments, while clearly being the backbone of support of industrial agriculture (half of farm income is from federal government support) are mostly irrelevant to organic farmers. At the same time, organic farmers are missing out on research dollars as well, since organic farming research has been relegated to the backseat and receives little financial support.

Organic farming has been playing catch-up to the conventional industrial research system since its modern inception in the 1970s. Industrial agriculture is strongly supported by the USDA and land grant university structures. The 1862 Morrill Act established the land grant colleges to "teach such branches of learning as are related to agriculture and the mechanical arts" and to promote "practical education of the industrial classes" (National Research Council 1996). This, with several later pieces of legislation, set up the three components of the land grant system: teaching, research, and extension (this provides public service to society through technology transfer and extended education). The basic tenant was that practical information was to be studied and brought to the average citizen. Unfortunately, most of the agricultural research that has been conducted now is of little value for organic farmers because the topics are irrelevant and much of the research is related to agrichemical studies. First, "research results from biologically impoverished conventional farming systems cannot be easily transferred to organic farming systems, since plant nutritional and resistance conditions and the biological environment have profound effects on disease management" (van Bruggen and Termorshuizen 2003, 154). Organic farms are more diverse agroecosystems, with healthier, balanced soils, so conventional research simply does not apply. Second, the core of agricultural research is controlled by agricultural input corporations, such as pesticide manufacturers, that provide significant funding for land grant university research on their particular chemical inputs. Obviously these corporations see no benefit in funding organic farming research that would lead to a reduction in the use of their agrichemicals. As noted in the Acres USA *Ecofarm Primer*, "The answer to pest crop destroyers is sound fertility management in terms of exchange capacity, pH modification, and scientific farming principles that USDA, Extension and Land Grant colleges have refused to teach ever since the great discovery was made that fossil fuel companies have grant money" (Walters and Fenzau 1996, xiii).

This lack of funding for organic methods influences the topics of university research, with organic techniques seldom being the subject of study. In fact, the general agricultural faculty mind-set has been in strong support of conventional industrial agriculture (Beus and Dunlap 1992). Some faculty doing research on nonindustrial modes of production note that institutional biases and links to agribusiness hinder the availability of information on alternative research (Larson and Duram 2000). Land grant universities have 885,863 acres of field plots and research lands in the United States, but only 0.02 percent (151 acres) are certified organic (Sooby 2003). This is one hundred times less than the 0.2 percent of total U.S. cropland that is certified organic and hundreds of times less than the percentage of some crops; for example, 2 percent of tomatoes grown in the United States are certified organic. The complex nature of agricultural policy makes a shift toward alternative research iffy, as researchers have had neither professional or financial incentives to move in that direction (Smith 1995). There are a few glimmers of hope with promising organic research programs under way at a few universities, but this historical lack of research attention shows the low status organic farming has had within the conventional industrial agricultural research system. At the same time, demand for organic food has been increasing by leaps and bounds.

Organic farmers have been on their own to experiment and develop pest management and soil fertility techniques and to share their findings with other organic farmers in their region or across the country. This is particularly an acute need for organic farmers, as information needs intensify with the adoption of reduced-chemical methods (Lockeretz 1991). Even programs in "sustainable" agriculture are often irrelevant for organic farmers. The most important national initiative for sustainable agriculture is the USDA's Sustainable Agriculture Research and Education (SARE) program. Yet, even within this targeted program, only 19 percent of the funds are for projects that focus on organic production and marketing (Greene and Kremen 2003). As we found in our study of researchers who were funded by the north central region of this grant program and organic farmers in the same region: there was a very little overlap between the desired topics that organic farmers found useful for their farm and the actual topics of the SARE research (Duram and Larson 2001). This varies by region, however, as the north east SARE program has funded conferences and information relevant to organic farmers (see Stoner 1998, for example). It is hoped that SARE is becoming more sensitive to organic growers' needs, as presented on their current Web site. Although even here, the term *organic* is somewhat taboo, even when the topic is appropriate. In an information bulletin titled "A Whole-Farm Approach to Managing Pests" the term *organic* is not used (USDA–SARE 2000). Apparently, *pesticide-free*, *ecological approaches*, and *integrated pest management* are the terms in favor.

In an overview study by the Organic Farming Research Foundation

(OFRF), Lipson (1997) looked at the research topics of projects that were funded by the USDA. Of course a search of *organic* was not possible because this term was not a recognized search variable. But Lipson used seventy-one related terms such as *compost* and *crop-rotation* in an attempt to find all research relevant to organic farming. Incredibly, of the thirty thousand USDA research studies, only thirty-four have a "strongly organic" focus, and these represent less 0.1 percent. This is a poor showing for an agency that admits organic crops and foods are one of the fastest growing segments of U.S. agriculture.

The organic taboo is an inexplicable phenomenon that goes back a few decades. The 1980 *Report and Recommendations on Organic Farming* was actually commissioned by the USDA in the late 1970s, but it was basically ignored by the agency (Youngberg et al. 1993). The term *sustainable* became the accepted term in policy arenas, and "organic" has been used as a label to designate a specific type of commodity (Klonsky and Tourte 1998). This is likely due to the fact that the term *sustainable* is acceptable within the industrial production mind-set. Perhaps because it is so nebulous, it never really challenges the industrial status quo. So "sustainable" is a safe term – and commonly used by environmentalists and agrichemical companies alike (Youngberg et al. 1993). In contrast, organic has a real meaning; and it requires a radical shift away from the industrial system, since synthetic agrichemicals are prohibited in organic farming. This leap requires, to some extent, a denunciation of the status quo, and obviously the USDA could not or would not take that step. Thus government programs inundate us with the vague term *sustainable agriculture*, which may in fact look very much like industrial production. It has been convenient to keep the term vague so that there is no real change to the conventional system.

There is confusion around this term *sustainable agriculture*. Its definition generally includes concepts of ecological, economic, and social sustainability (USDA–SARE 1998). But it is fuzzy; even the USDA's Sustainable Agriculture Research and Education (USDA) program provides a confusing definition: "sustainable agriculture encompasses broad goals, and farmers and ranchers develop specific strategies for achieving them" (USDA–SARE 2003, 2). Indeed, most farmers would say they are sustainable – who would proudly admit to unsustainable use of resources? Compounding the problem is that farmers themselves don't really know what sustainable means, as one study shows most farmers think it relates only to environmental concerns and not economic or social issues (den Biggelaar and Suvedi 2000). And the term can be misused, as it is "embraced by virtually every constituency with an interest in agriculture" from environmentalists to farm input manufacturers (Youngberg et al. 1993, 295). Sadly, the use of the term *sustainable* is so popular that it has no "unifying vision" and is, to some extent, "meaning *less*" (Marshall 2000, 268). All of this is confusing for consumers. What is a "sustainably produced" crop? Is it worth paying more for a "sustainable" apple if we don't really know what that word means?

On the other hand, the term *certified organic* is a real, definable term for consumers. We can define it because it is based on a specific process of certification. Farmers must forego synthetic agrichemicals for three consecutive years; they must maintain detailed farm histories; they must document every input to their fields; they must have an annual inspection by an outside inspector; and they must show that they are building their soil through rotation and use of green manure (crops planted and plowed under to fertilize the soil).

The misuse of the term *sustainable* by some is unfortunate, as there are so many valuable activities undertaken by honest proponents of sustainable agriculture. For example, the National Campaign for Sustainable Agriculture (2003) describes their mission as "educating the public on the importance of a sustainable food and agriculture system that is economically viable, environmentally sound, socially just, and humane." They post very clear descriptions that help farmers and others understand the status of specific national agricultural programs. They also "help grassroots concerns and priorities be heard in Washington DC," which is a monumental task, indeed, given what we're up against.

But perhaps the sustainable agriculture movement has an opportunity to take back the meaning of sustainable – and protect genuine organic methods at the same time. If sustainable agriculture could rally behind certified organic farming, it could help consumers understand what they are buying, and it could provide attention to the movement to keep organic farming "honest" in terms of size, operation, and ownership. Thus, if certified organic farming could become the accepted method of "sustainable" agricultural production (thus providing an absolute definition of organic for consumers, environmentalists, and policymakers), then organic farming could defend itself against some of the negative forces stunting its honest development today.

ORGANIC FARMING TODAY

We are at a crossroads. Organic agriculture has matured beyond its hippie roots of the 1970s garden patches. Today organic farms are large and efficient. They provide obvious alternatives to the economic and environmental downfalls of industrial agriculture: Farmers can create a new marketing system for themselves that is outside the large, industrial farming system, and clearly certified organic farms do not employ the synthetic pesticides and fertilizers that are mainstays on conventional farms. These farms are resilient, as organic farmers adapt to change, act independently, and learn new techniques as needed (Milestad and Darnhofer 2003). So organic farming is better for the environment (Xie et al. 2003) and better for small family farmers.

Buying organic food is not only increasingly trendy but truly good, both for health seekers and environmentally conscious shoppers. Consumers are aware of organic products and willing to pay a bit more to buy these items. The drawback to this rapid growth in demand is that organic farming can sometimes be pulled into the industrial mode of production, which I nickname "Big O Ag." Of course, agribusiness corporations realize the large market growth in organic products. Consumers now seek organic cereal, cookies, canned goods, and produce, and food companies are well aware of these trends. There is concern that this is a time of transition for organic farming, distribution, and retailing, as it is getting bigger and thus more similar to a conventional food system (Dimitri and Richman 2000). For example, Tree of Life and United Natural Foods (which bought out Blooming Prairie and Northeast Cooperatives in 2002) are the only national distributors of organic foods, controlling 80 percent of the market. In addition, "Whole Foods, Trader Joe's, and Wild Oats together have over 440 stores across the United States and sell approximately $5.5 billion in natural and organic foods each year," which "gives these chains a dominant presence" in retailing (Sligh and Christman 2003, 26). If there is a market for millions of one-pound bags of precut organic carrots, then a large farming corporation will step in to produce these carrot bags and ship them across the continent, with little regard for the "buy locally" philosophy. Likewise, organic milk has become dominated by a single corporation (Horizon Organic Dairy was recently bought out by Dean Foods) that uses ultra-pasteurizing techniques to lengthen shelf life so that the products can be shipped across the country. The huge dairy corporation claims to rely on smaller family farms for their milk supplies, but the scale of this corporation is in sharp contrast to the image many consumers have of organic production. How will these consolidations affect the organic products that consumers demand and the crops that organic farmers grow?

This is not a uniform shift within organic farming, all toward Big O Ag. Rather, separate farming sectors are influenced differently. The organic produce sector seems to be most readily assimilated in the industrial mode of agricultural production, perhaps because of the long history of California produce production and the early interest in organic crops in that state. They have had time to practice and hone their large-scale organic techniques, which often resemble their nonorganic counterparts (minus the synthetic chemicals, of course). These farms are not the small family-operated organic farm in our hypothetical mind's eye; rather, these are large produce operations that employ many migrant workers and have sophisticated national or international marketing channels. Large-scale organic grain farms also sell to national or international markets, although they remain more family run, due to lower labor demands than the produce operations.

The small and medium-sized family organic farms also exist, and organic methods often provide these farms with a genuine opportunity that would not be realized in conventional production. These family organic farms are seen in various forms in all regions of the United States. These farms rely on farmers' markets or other types of direct marketing such as Community Supported Agriculture (CSA). Another type of direct marketing, present among specialized crops such as citrus, is mail-order sales. Medium-scale family organic farms are found in the plains and prairies, and these operations are generally mixed with both grains and livestock. These farms look and feel like what most of us consider a traditional family farm (but which sadly rarely exists today). These small and medium-sized organic farms are still true to the classic organic image: family owned, locally marketed.

Organic Farmers and Geography

The geographic distribution of organic cropland varies from the patterns of industrial production. For example, 80 percent of U.S. industrial cropland is in just four crops: corn, wheat, hay, and soybeans. But these crops represent only 49 percent of certified organic cropland (Klonsky 2000). On the other hand, organic vegetables are grown on 12 percent of certified organic cropland, compared with only 1 percent of total U.S. cropland. These cropping variations are seen in regional differences as well, with the Pacific and Mountain regions offering two-thirds of organic cropland but only one-third of total U.S. cropland. The opposite is seen in the Corn Belt, which contains only 11 percent of certified cropland but 25 percent of total U.S. cropland. Thus the vegetable growers in the western United States are a significant part of organic production. Although there are geographical variations in organic production, there are some similarities across regions.

The common thread that exists on all successful family organic farms is the farmer's willingness to try new crops, farming methods, or marketing locations. These farmers seem to meet the adversities of American industrial agriculture head on; they have chosen a distinct path and are proud of it. They are quick to point out the differences between their operations and their conventional neighbors: better soil quality, more crop diversification, lower debt, and unique markets. Organic farmers often show visitors that the soil on their farm is rich from organic compost and is less compacted. They grow more types of crops and more diverse crops that are not common in their region. It's a challenge to find reliable, distinct organic marketing avenues that will provide the required price premiums on all the various crops they grow. Many organic farmers do their own on-farm experiments to compensate for the dearth of relevant government- or university-sponsored research. And most contact one another, or indeed anyone, who may have information.

Many studies have investigated the characteristics of organic farmers. There are, it seems, as many similarities as differences. On-farm research comparisons help us understand how organic farms "look" and function.

2

The Science of Organic Farming

Ecologically, agriculture is a highly effective means of converting solar energy into food and fiber. Given sufficient water, and properly managed, the system can operate provided with nothing more than sunshine. But modern agricultural technology has disrupted this efficient relationship.

– Barry Commoner, *Making Peace with the Planet* (1992)

arly studies of organic farming faced the obstacle of definition: since organic certification was not readily available, it was difficult to identify organic farms. This meant that it was impossible to obtain a list of organic farmers to study; so researchers had to use less formal means to identify them. As recently as the early 1980s, "very little research has been done on organic farming as a system of commercial agriculture" (Lockeretz et al. 1981, 541). At this time, many people still assumed that organic farming was just a throwback to methods from the 1800s, although research into the social and economic forces at play on organic farms was beginning to show otherwise (Lockeretz et al. 1978).

This reminds me of an elderly conventional farmer who grows chemically intensive vegetables in Colorado. When I asked him about organic methods, he huffed, "Huh! I remember that from when I was a boy. It means worms in your cabbage!" Well, we've come a long way, partly because of innovative organic farmers who've experimented and learned through trial and error and partly because of the relatively scarce research that helped educate people about modern organic farming. Organic farming techniques are increasingly sophisticated, with improved cultivation and tillage implements, new crop varieties, and better management knowledge.

Arguably "the first farm-scale study of modern commercial organic farming in the United States" was conducted in the mid-1970s and investigated crop yields and production costs. These researchers realized that "we cannot predict the performance of modern organic farms simply on the basis of yields obtained three decades ago or from older historical plots" (Lockeretz et al. 1978, 130). So they set out to compare fourteen pairs of organic and conventional farms in the Midwest and found that even with no organic price premiums, organic farms' economic performance was equal to that of conventional farms over three growing seasons, 1974, 1975, and 1976. In addition, the organic farms had the benefit of less dependence on off-farm inputs, fewer external energy supplies, less soil erosion, and higher soil organic matter. This early study must have been quite a surprise for many people who considered organic farming a hippie throwback to preindustrial times. It actually showed that commercial organic production was economically viable and offered advantages over conventional farming.

More work was published a few years later by the same authors (Lockeretz et al. 1981). This time they investigated both the social context and the field methods employed by organic farmers in the Corn Belt region. Mailed questionnaires from 174 organic farmers indicated that 80 percent of them had started in conventional farming, as opposed to being newcomers to agriculture. Their most common reason for shifting to organic methods was concern about chemical use (including health of humans, livestock, and the soil and the ineffectiveness of the chemicals). The farmers also described clear barriers to organic methods: difficulty finding markets, lack of information sources, weed problems, and being shunned by conventional farmers. The farming practices of 363 organic farmers in the region were generally similar to conventional farmers, minus the agrichemical use. Instead, they employed crop rotation (planting different types of crops in a specific order) for insect control and fertility. Farm types included a mix of crops and livestock, so manure could be applied to build soil fertility. Organic farmers tended to have a higher number of crops in their rotations. But overall, this early research disproves the stereotypes of that time and shows that organic farms were not drastically different from their conventional neighbors.

Ten years later, a follow-up study showed that most of the original group of farmers was still in business and still using organic methods (Lockeretz and Madden 1987). These farmers had sound financial standing with very low debt compared with conventional averages in the region. This is particularly remarkable given the severely depressed Midwestern farm economy of the time. The farmers still mentioned the health benefits of organic farming and felt there was little change in the lack of institutional support for organic production. These initial studies paved the way for other research of organic methods. Today we see how far organic farming has come – and realize the challenges it still faces.

COMPARING ORGANIC AND CONVENTIONAL FARMS

What methods should be used to compare organic farms to their conventional counterparts? This is a complicated question. The long-term management actions of experienced organic farmers who are intimately familiar with their own land would certainly provide different results than an agricultural plot on a university research station in which researchers simply remove the chemicals and attempt to begin organic production on one field. While agricultural research is most often conducted under these controlled conditions, organic methods show different results when studied holistically – as a whole operating farm, rather than just one field under a specific one- or two-year study. Then the issue of equitable comparison arises: can two neighboring farms be identified that provide examples of organic and conventional methods on similar soils, with similar crop types? Organic farms tend to produce a larger number of distinct types of crops that are not typical of conventional farms. Still there is comparative research that addresses such concerns.

One comprehensive survey of organic agricultural productivity was especially thought provoking because it described the various types of comparisons that are possible (Stanhill 1990). Specifically, organic farming demands a holistic research approach to reveal the inherent symbiotic relationships that occur, and it is difficult for the reductionist methods of scientific inquiry to capture this complexity. Thus the author reviewed many types of historical studies (some dating back to the 1940s) to draw conclusions about organic productivity. This included an overview of 205 yield comparisons for twenty-six crops and for milk and eggs from data gathered at fifteen sites in North America and Europe. Data from comparative observations of operating commercial farms, long-term replicated field plots studies, and whole-system experiments were included. First, in thirty comparable observations of active organic and conventional farms, organic crop yields exceeded conventional in thirteen, were equal in two, and were less in fifteen. Second, in 104 crop seasons of controlled field experiments, organic plot yields were lower than conventional 75 percent of the time (but only half of these were statistically significant.). Attempts to study whole systems over the long term run into the problem of consistency. Even in one location, for example, results were mixed because the management methods had changed so much over the 1938–75 trial period. Here, out of sixty-six comparisons, organic yields were lower two-thirds of the time; of those, most of the yield differences were less than 20 percent. Overall, the studies showed that organic crops outyielded conventional crops in one-third of the cases, and organic crops generally outperformed conventional crops in adverse growing conditions. So, taking into account all the systems, crops, and methods, organic yields were mostly within 10 percent of conventional yields, and they achieved this with no agrichemicals and relatively little research information to guide farm management.

A very important article on the long-term productivity of organic farming was published in *Science* in 2002. Findings were based on twenty-one years of comparative data from conventional and organic farming systems with identical crop rotation, varieties, and tillage in Central Europe (Mäder et al. 2002). In terms of agricultural sustainability: "We found crop yields to be 20 percent lower in the organic systems, although input of fertilizer and energy was reduced by 34 to 53 percent and pesticide input by 97 percent. Enhanced soil fertility and higher biodiversity found in organic plots may render these systems less dependent on external inputs" (1694). Reading this concise article, we are encouraged by the authors' suggestions that appropriate plant breeding may further improve yields of some organic crops. Just imagine how far organic methods could advance if we had research dollars pouring into these topics. Their soil studies confirm that organic fields had greater soil stability, microbial diversity, microbial activity, earthworm biomass, predator insects, weed diversity, and more completely decomposed plant materials. In a sense, this long-term organic farming system is similar to a mature natural system due to its diverse fauna and flora and efficient resource utilization. Mäder et al. conclude, "Organically manured, legume-based crop rotations utilizing organic fertilizers from the farm itself are a realistic alternative to conventional farming systems" (1697).

Farming with organic methods can now produce yields similar to convention methods, and if organic farming methods could obtain agronomic research funding, these yields would be even greater in the future. While crop yields are a favorite point of comparison, farms only exist if they are economically viable. Regardless of how much of a given crop is produced, organic farmers may be more likely to turn a profit because they don't have to pay high agrichemical bills. Plus, the economic benefits of selling organic crops at a price premium often outweigh the slightly lower yields, thus farmers can be profitable without being caught up on the industrial treadmill of production.

Detailed Regional Assessments

Comparisons between conventional and organic farming methods often take a distinctly regional approach in the United States. In the plains, an economic analysis of several cropping systems in Nebraska found that use of organic versus conventional methods "had little impact on profitability" (Helmers et al. 1986, 153). Four cropping systems (continuous row crop, rotational row crops, rotation with small grain, and organic) were compared over an eight-year period, and organic methods performed well in terms of profit and risk, even without including the benefits of organic price premiums. Even when sold at conventional prices, the organic production system had similar profit margins to the conventional fields.

Also in the plains, the economic potential of organic was comparable to conventional and ridge tillage (a soil conservation method) cropping in South Dakota (Dobbs et al. 1988). Again, no organic price premiums were assumed, and organic yields were somewhat lower than the agrichemical crops (particularly with corn, while soybean and wheat yields were more similar). Due to lower direct costs, the organic system was competitive. In addition, the researchers modeled various scenarios and found that with reduced federal farm program levels, the organic system would fare better (because most organic systems do not rely on such subsidies). Continuing with this research, the crop yields and economic performance of organic, conventional, and reduced-tillage farming systems were compared over a five-year transition period at a South Dakota experiment station (Smolik and Dobbs 1991). Wheat and soybean yields were similar among the systems. The five-year net incomes were highest for the organic system because direct production costs were lowest and crops sold at premium prices. In addition, the organic yields were more reliable than conventional crops in drought conditions. Other research concurred that under drought conditions organic systems have an advantage (Rickerl and Smolik 1990). Even on newly transitioned fields, which would not yet exhibit the cumulative effect of many years of organic soil building, soybean yields were high and, as mentioned, the organic methods proved beneficial through the variable precipitation on the plains.

In Ohio, as in most states, there is plenty of data on conventional farming but little on organic methods. So the profitability of organic and conventional farms was compared by gathering information from a mail questionnaire of sixty-four organic farmers and using the state's longitudinal data gathered on conventional farmers (Batte et al. 1993). Three factors showed the most significant variation: organic farmers received price premiums for their crops, organic yields were lower than conventional, and organic farms paid much less for fertilizers and pesticides. Thus profits were slightly higher for the organic farms. Since, on average, organic farms (181 acres) are smaller than conventional (235 acres), the authors noted that if organic methods could be transferred to larger acreages, the organic profit margins would in fact be even higher. Many comparisons are actually slanted by the fact that "average" organic farms are smaller than "average" conventional farms. Perhaps a more accurate comparison would be to use the organic farm size and just a portion of the conventional farm (to avoid skewing the results toward larger farmers). This indicates that organic farmers can succeed with relatively fewer acres than is necessary in the high-yield mentality of conventional farms.

Farming systems data from the Rodale Institute in southeastern Pennsylvania showed that management of an organic system evolves over time, as a manager becomes more knowledgeable about the farming conditions (Hanson et al. 1997). This study compared three different (complicated and evolving) organic rotations with a typical conventional corn-soybean system. The early years in the organic rotations exemplified the difficult transition years – coming off the chemicals, but not yet having the soil in organic equilibrium. Corn yields during the transition to organic were 29 percent lower than conventional yields, but then rebounded to an average of just 2 percent lower than conventional yields. Net returns (figured as revenue minus explicit costs) were substantially higher for the organic system. However, when family labor costs are figured as dollar amounts (which most farmers do not do – they simply work on the farm as much as they need to), labor would make the net returns equal to or lower for the organic system. Thus this study showed that farmers have to work more hours to take care of an organic crop farm than a conventional one, and this variation was especially pronounced as farm size increased. The time factor makes large-scale part-time organic farming nearly impossible, while it is quite common for conventional farmers. Organic "per-acre returns are competitive and sometime greater than conventional grain rotations," but they "required much more family labor," and this hinders a "farmer's ability to continue farming while working full-time off farm" (8). This, however, begs the question: why should farmers have to work off farm? Why can't they earn enough from farming to make a living? As I discussed in the first chapter, the industrial agricultural system is forcing farmers off the land or into farming as a hobby. Organic farming allows farmers to earn a fair price per bushel, which can allow them to work only one job – on the farm.

Iowa established a long-term agricultural research site in 1998. Based on three years of comparative data, results show that organic farming is more profitable than conventional production (Delate et al. 2003). A conventional corn-soybean rotation was compared with two organic rotations: corn-soybean-oat and corn-soybean-oat-alfalfa. In each case, the organic and conventional yields were similar, but the input costs of organic farming were lower. Overall, the organic methods proved to be substantially more profitable than the conventional methods.

Dealing with similar crops, the Illinois Stewardship Alliance published a report on their farming systems study that compared the economics of conventional, no-till, three-crop, and organic methods on five hundred acres of adjacent fields in east-central Illinois (2002). The study ran for six years and provided a wealth of information and examples about the complexity of comparisons. Comparing the average annual net return per acre, the organic system earned the highest in five of the six years and had some of the lowest average costs. Organic crops earned price premiums, and organic methods kept input costs low. The organic crop rotations were diverse: blue corn, white corn, wheat, soybeans, and rye cover crop. The cost figured for labor in the organic system decreased over the six-year study, which underscores how farmers often gain more knowledge about organic management through experience (or, as the report notes, perhaps the farmer became somewhat more tolerant of weed occurrence). Overall, these studies in Iowa and Illinois showed that organic methods are quite competitive with conventional methods in the Corn Belt. This surprised many people in this region where (conventional) corn is king (Pollan 2002).

A specific study of corn roots was based on samples from low-input and conventional farms and found that root length density was significantly greater in the organically fertilized corn (Pallant et al. 1997). It follows that the greater root density would allow corn to absorb more nutrients and water, even in drought conditions. This may explain why organic corn yields tend to be more stable in adverse growing conditions, while conventional crops fluctuate greatly. We may also question how this could affect us when the global climate changes. Although we can only theorize now, it is possible that currently fertile areas may become significantly warmer. Organic methods, if fine-tuned, could help us withstand these adverse growing conditions.

A comparative study of economic and yield data from California crop production was based on eight years of information (1989–96) from a farming systems project in the Sacramento Valley that included conventional (four- and two-year rotations), low-input, and organic management (Clark et al. 1999). The key factor in profitability was the frequency of tomato cropping in the rotation, as this crop earned high prices. The organic system was the most profitable of the four-year rotations because of organic price premiums. Problems in the organic system were related to high nitrogen demand and weed management costs. This study exemplified the complexity of crop comparisons, as the organic system had a complicated rotation of crops that built soil fertility: cover crop, safflower, cover crop, corn, oats/vetch, bean, cover crop, tomato. The two-year conventional rotation was simply tomato, wheat, tomato, wheat (and soil fertility was supplied by agrichemicals). Given the profitability of tomatoes, it is clear why some farmers would choose this two-year rotation. Organic farms must be guided by longer-term goals and ecological balance. And the authors note that this study did not include the long-term benefits of increased soil organic matter, nutrient storage, or reduced erosion.

The Sustainable Agriculture Farming System Project in the Sacramento Valley provided information to study the long-term effects of organic, low-chemical input and conventional farm management (Colla et al. 2002). The data was detailed: From 1989 to 2000 the specific field methods such as planting dates and input names were logged. The soil chemical properties, crop yields, and crop mineral compositions were compared. Results showed that the organic soils developed the highest levels of carbon, nitrogen, soluble phosphorus, exchangeable calcium, and potassium. Organic tomato yields were similar to the conventional and low-chemical input yields. Organic tomatoes contained higher levels of phosphorus and calcium, while conventional tomatoes had higher levels of nitrogen and sodium. While the levels may not lead to human health concerns, the 38 percent higher nitrate levels in conventional tomatoes would likely cause metal storage containers to corrode. Over the long term, organic practices produced tomato yields similar to conventional farming methods, with richer soil and more minerals in the food crop.

The "purported drawbacks" of organic tomato farming that "include an increased incidence of pest damage and higher risk of pest outbreaks" were investigated in California (Letourneau and Goldstein 2001, 557). This fascinating study compared tomato production on eighteen commercial farms (half were certified organic) and found that damage to tomato foliage and fruit did not differ between certified organic and conventional farms. Specifically, this research focused on arthropod communities (that means "bugs" to most of us: spiders, beetles, ants, etc.). Both types of farms had arthropod damage, but there were big differences in the community structures – organic farms had higher richness of species and natural enemies. This indicated that any one particular pest bug would be "diluted" and have more potential predators if it were in an organic tomato crop. The abundance of bugs was more associated with specific on-farm tasks and landscapes (fallow fields, surrounding habitat, and transplant date) than with whether the place was organic. This refutes the commonly purported view of organic farming – that pests destroy much of the crop. That is simply not true. Now imagine what we could accomplish if we had millions of research dollars to study cutting-edge organic farming techniques.

A comparative study of organic, conventional, and integrated (combination of organic and conventional) apple orchard systems was conducted in Washington state between 1994 and 1999 (Reganold et al. 2001). Yields were similar among all three, and there were similar low levels of fruit damage, pests, and diseases. With organic and integrated methods, soil quality was higher in terms of accommodating water availability and supporting fruit quality and productivity, and environmental impact was lower in terms of chemicals used and their associated impact ratings. Organic apples tasted sweetest, were most energy efficient, and had the highest profitability.

Another comprehensive study of apple production was conducted between 1989 and 1991 in California (Swezey et al. 1998). Here a conventional orchard was compared with an organic orchard during its three-year transition. The organic system had higher apple tonnage (due to hand thinning versus chemical thinning) and higher number and weight of fruit per tree (but smaller average fruit size). Price premiums were approximately 35 percent for certified organic apples, so even with higher material and labor inputs, the organic system earned greater net return per hectare. There was no difference in pest damage between the two orchard methods, although the organic trees relied on pheromone-based mating disruption while the conventional acres used synthetic sprays. The presence of weeds was greater on the organic fields, but this did not decrease yields. Earthworm biomass and abundance increased by the third year, as the organic methods began to improve soil quality. Overall, the organic apple production system was successful in its three years of transition and would likely remain successful as the farmers became more experienced with the specific methods and biological interactions involved. Why continue with conventional apple production if organic apple farms produce better fruit and earn more profit?

In addition to comparative crop studies, research has also been done comparing conventional and organic dairy production. A comparative study of the economics of milk production was recently conducted in California (Butler 2002). This research indicated that the cost of production, figured per cow, was 20 percent higher for organic dairy farms. Although there is a price premium, this did not offset the cost discrepancies, so net income per cow for organic producers was only 75 percent of that for conventional producers. This paints a relatively bleak picture of the organic dairy industry, which is increasingly dominated by a few larger dairies (Pollan 2001b). Yet we must consider the long-term prospects for organic milk production. Perhaps with increased consumer demand and more experienced organic dairy farmers, these income discrepancies can be leveled. For example, with increased consumer demand, more processing plants may cater to organic milk. Thus transportation costs would be reduced if farmers didn't have to truck their milk so far. Also, dairy farmer cooperatives are successful in some parts of the country.

The health of organic versus conventional dairy herds was investigated through paired herds of similar size and location (Hardeng and Edge 2001). Data from thirty-one organic and ninety-three conventional dairy herds were gathered between 1994 and 1997. Organic milk yields were approximately 20 percent lower than conventional, but the organic cows showed substantially better health, with only one-third as many cases of mastitis and ketosis. The cows were older in organic herds, indicating that they produce milk longer than the conventional cows. These health benefits may be due to increased time outdoors and superior (organic) feed.

Turning to an international perspective, an overview of organic farming in Australia outlined the geographical factors that influence organic farming "down under" (Conacher and Conacher 1998). Due to vast distances and low population density, farms were often isolated, which affected marketing options. This is also an issue in many parts of the rural United States. Likewise, climate variability and poor soil quality can impact farmers in some areas. On a more positive note, the authors noted the environmental benefits of organic production, with a small warning about possible negative effects. (Of course, biological pest controls are not completely benign.) Overall, organic farming can be economically and ecologically beneficial.

Comparative Models

Not all agricultural research involves visiting a farm; some researchers opt to use statistical data and build hypothetical computer models to simulate long-term agricultural conditions. For example, the energy use and economic outcomes of conventional and organic production of maize and potato were estimated through statistical modeling (Pimentel 1993). The relevant data were general estimates of labor demands, fuel use, yield, and input costs, but provided no regional or local verification on actual farms. The model estimated that organic maize would have lower costs of production with higher yields and less fossil fuel input, compared with conventional production. While indicating some of the important variables in comparing farming systems, this study omits many important geographic considerations such as location, soil type, crop rotation, and specific management techniques, which could be added to tailor such a model to a given location.

Also dealing with energy use, a hypothetical model of fossil fuel use on conventional and organic farms was developed in Denmark (Dalgaard et al. 2001). Both the direct (fuels, electricity) and indirect (fertilizers, pesticides, machinery, import of fodder) on-farm energy use were simulated. The model indicated that energy use was lower on organic farms, making this system more energy efficient (both for crop and livestock production). Various scenarios were created: differing levels of imported fodder, production of pigs (lower efficiency), and production of cattle (higher energy efficiency). Overall, there was a relationship between lower energy use and lower yield. Energy use and per unit crops and animals produced was consistently lower in the scenarios for organic farming, which implies greater sustainability. This study only considered energy use in production of the crops – not transporting the crops after harvest, which would add another dimension to the issue of energy use.

Using data from the USDA–Agricultural Research Service's Sustainable Agricultural Demonstration site in Beltsville, Maryland, researchers created a model to simulate sixty years of cropping (Lu et al. 2003). Six types of grain cropping systems were modeled, including a two-year organic rotation (which is very uncommon in certified organic production) and a three-year rotation (this and a four-year rotation are more common in certified organic production). The organic systems had lower profit variability, low erosion risks, and no risk of herbicide contamination compared with conventional methods. Overall, this simulation model indicates that organic methods may be quite good for farmers seeking to avoid economic and agrichemical risk.

Another study that presented a model, as well as describing more theoretical aspects of the organic/conventional comparison, was conducted in Greece (Tzouvelekas et al. 2001). Here the technical efficiency of approximately eighty-five pairs of organic and conventional olive farms was assessed by statistically estimating efficiency in the production of one kilogram of olive oil. The factors considered were percent of land in production, labor, input costs, and other costs. Also the inefficiency effects model was based on family labor percentage, farm size, capital inputs, and environmental variations. In either case, organic olive farms had a higher degree of technical efficiency in relation to their production level. The efficiency and energy models indicate that organic farms have the potential to endure well into the future.

Comparing Soils

Soils are of higher quality on organic farms than on conventional farms, exemplifying the major difference between these two farming methods. Organic farmers work to build their soil through crop rotations, improving it for the next generation. Conventional farms tend to use the soil as a medium to hold the seed, applying the necessary chemicals for the plant to grow. For example, a Dutch study found substantial differences between the soils from farms that produced similar crops under organic and conventional management for seventy years (Pulleman et al. 2003). On these farms, the soil organic matter content, the earthworm activity, and mineralization were all greater on organically farmed soils. Further, the organic soils had higher water-stable aggregation, which indicates that they hold together when wet and would be less susceptible to erosion. So there are "beneficial effects of organic farming" related to both the soil biochemical properties and "soil physical aspects" (157).

In a U.S. study, the soil structure was distinctly different on two adjacent farms in eastern Iowa: a 35.6 acre (14.4 ha) conventional grain farm and a 4-acre (1.6 ha) organic vegetable and grain farm (Gerhardt 1997). Both farms had been in operation since the 1950s, so results indicated the long-term variations in field methods on silt loam soils. Soil samples gathered from five sites on each farm underwent multiple tests. Results showed that the organic farm had significantly deeper topsoil with coarser texture and higher organic matter, porosity, and earthworm abundance. It did not show signs of compaction and erosion, which were obvious on the conventional farm. Crop types varied considerably between the two, and variations in cropping and tillage affect soil structure. We can conclude that the management methods associated with organic farming (e.g., crop rotations, organic inputs) help maintain and improve soil structure, which will provide better crop growth.

Variations between how organic matter breaks down in soils on conventional and organic farms was studied in Ohio (Vazquez et al. 2003). Overall, earthworm population density was much higher in organic soils. Decomposition was higher on the organic farm, where more soil decomposers are active, particularly in warm weather. This suggested that "a more active soil biota" existed on organic farms (559), which was due to management techniques. A study in the Netherlands echoed this concept, as researchers found the "bacterial biomass occurring under organic farming scores higher than in other farming systems" (Mulder et al. 2003, 516). And German researchers found that soils on organic farms have much higher infiltration rates than soils found on conventional farms, and this capacity helped protect nearby areas from flooding (Schnug and Haneklaus 2002).

Variations in soil quality were found in Denmark between conventional and organic farms on humid sandy loam soils (Schjønning et al. 2002). The conventional and organic dairy farms were "integrated" grain/cattle systems, which is not the case on many large-scale conventional dairy farms in the United States and elsewhere. Overall, the use of heavy machinery caused compaction in both farming systems. Beyond that, organic manures and diversified crop rotations did improve soil quality, specifically leading to higher microbial biomass carbon on the organic farms. Back in the United States, five pairs of organic and conventional farms with matching soil types were investigated (Liebig and Doran 1998). The organic farms had healthier soils, with more organic carbon and total nitrogen (with higher ratios of mineralizable nitrogen to soil nitrate), lower bulk density, and better water-holding capacity. Organic methods have the ability to improve soil quality due to their use of diverse crop rotations.

A review of several previous soils studies found that in nearly every instance organic farms had higher levels of soil organic matter than did conventional farms (Shepherd et al. 2002). In addition, these studies found that organic farms tend to have deeper topsoil with less erosion, more granular structure, more friable consistency, darker color, and more active earthworm populations. These authors noted that better soil structures were present when fresh organic residues were added often – which occurred on organic farms. But, they say, if conventional farms would consistently add this quality organic matter to their soils, they would also have higher soil organic matter content. So organic farms do the right thing, as part of their normal management approach, but conventional farms don't because it doesn't fit into their system of farming – one that is built on growing a few crops and relying on chemical applications rather than crop rotation and organic matter for fertility.

While past studies and comparative studies are important – proving the viability of organic production over conventional methods – in the future we need research funding to focus on specific methods that would benefit organic production itself.

Organic Methods

It is important to be both realistic and practical when describing organic crop production. The transition to organic farming methods brings potential challenges that can be overcome by taking specific actions (Zinati 2002). Implementation of pest management procedures and the decision to grow specific crops depends on geography – the location, climate, resources, and past land uses. Farmers must work with their soils by balancing the organic and mineral components to increase natural controls that keep pests in check. Specifically, crop rotation, cover crops, mulches, crop diversification, resistant varieties, and mechanical cultivation are appropriate methods to build the soil. Pest populations must be kept at an "economically acceptable" level (606). Note this does not mean complete annihilation of all pests! The transition from conventional to organic production requires a "fundamental change in the farm operation" (609). There is no halfway option. Luckily, practical research such as this exists, providing specific information to assist farmers making the organic transformation. Here are the key points: use crop rotations to reduce pests and build the soil, change planting dates to break pest and weed cycles, don't let annual weeds go to seed, and increase crop seeding density so weeds don't have room to emerge.

Organic farming practices are complex and integrated (Watson et al. 2002). Field techniques are interrelated so that one action affects other components immediately and over the long term. It is difficult to view soil fertility in isolation from the environmental and production components of the farm. For example, crop rotations break pest cycles and with legumes can increase nitrogen availability to crops. Addition of manures and crop residues help recycle nutrients and influence soil organic matter, while short-term fallow fields also enhance organic matter and build soil structure and biological activity. Plus, there are ecological and economic consequences for nearly all field management actions. Certain changes occur when a conventional farm converts to organic methods (Langer 2002). A more diverse cropping mix is seen after the conversion, including more soil-building crops and fallow grassland.

Most modern varieties of seeds have been bred to perform under conditions of high synthetic chemical inputs (Watson et al. 2002). Thus, these crop varieties are not really appropriate for organic farming techniques. The fact that farmers only have access to conventional crop varieties is not commonly discussed in comparative articles. For example, one study found that organic wheat yield was 77 percent and barley yield was 74 percent of conventional yields (Entz et al. 2001). This makes sense considering that wheat and barley have been bred intensively within high input systems (Watson et al. 2002). But this was not discussed in the comparison of the crop yield and soil nutrient status of fourteen organic farms in the northeastern Great Plains, and huge variations in organic yields (from half to double conventional yields) were found depending on the crops (Entz et al. 2001). In addition, soil nutrients were sufficient, but the study did not include substantial analysis of soil quality (e.g., comparing soil organic matter). This is a good example of how modern science demands the reduction of a complex organic farming system into separate component parts that are not able to fully illustrate on-farm holistic interactions.

On the other hand, a multidisciplinary approach was employed to evaluate the ecological and economic aspects of one case study farm during its transition from conventional to organic methods (Cobb et al. 1999). The case study was in the UK, and many of the policy examples were more relevant within the European context, where government support payments for agroenvironmental activities and for organic transitions are more common than in the United States. Still, the research showed that organic farms were efficient in nutrient cycling, provided increased biodiversity, and improved soil health, and these ecological factors provided real social gains. So the authors make the case for permanent government supports for organic farming. They noted that the transition years were economically grueling, but the organic farm was profitable once the system was in place. Thus they recommended government programs and financial support to encourage the transition to organic systems, which benefit rural areas.

This raises the idea of input substitution (Rosset and Altieri 1997) in which organic farms increasingly rely on purchased off-farm inputs. This is a complicated issue because we can see how an integrated system of fertility and pest management is possible on a small scale. One farmer can handle twenty-five cows who produce manure that is composted and applied to twenty-five acres of cropland. But can a farmer – even an overworked organic farmer – make it economically with just twenty-five cows and twenty-five acres? Perhaps he can with multiple local direct marketing avenues and value-added milk-based products, but it would be challenging. Often a larger scale operation is needed, at least fifty cattle and two hundred acres, but then more labor is required, and that may mean hiring workers, which also takes away from the idea of a self-supporting integrated farm. So how can an organic farm be competitive and survive in modern American agriculture and also remain as a family-based operation? It seems that input substitution may be necessary at some level – perhaps so that farmers can concentrate only on growing and marketing crops minus the livestock.

This certainly doesn't seem to be an integrated system, and the purchase of off-farm compost may be necessary. I don't advocate a "bigger is better" viewpoint, but it is clear how it could happen.

ORGANIC LANDSCAPES

Aesthetic values are associated with farming, and many researchers have carefully studied the various factors that make a rural landscape diverse and colorful, visually coherent, and harmonious. Not surprisingly, organic farms tend to provide better landscapes and local environments than conventional farms.

Research investigated landscape features among seven organic and eight conventional farms in the Netherlands, Germany, and Sweden (van Mansvelt et al. 1998). The farm sizes ranged from 15 to 457 acres (6 to 185 ha) on farms that were established between 1947 and 1989, and utilized between two and sixty workers. Interviews, farm visits, environmental maps, photographs, and transects were all used in this analysis. A landscape should balance order versus wild nature and simplification versus holism, and this study found that organic farms exhibited more land use types, natural elements, crop rotations, woody elements, and farmyard variation. Diversity was greater on organic farms in terms of land use, crops, livestock, trees, flora, sensory elements (colors, smells, sounds), and labor. Yet there was coherence between the land use, local conditions, and spatial structures like field division and fences. True organic landscapes should not rely only on farm-level analysis; they must also include regional planning and cooperation. The authors suggest that a dozen or so neighboring organic farms could work together to reach broader landscape goals of diversity and a sustainable rural landscape. These are interesting ideas, but we must consider the differences between the United States and European rural landscape in terms of scale, farm size, and regional planning. The smaller land holdings and higher acceptance of regional rural/urban planning in Europe make the goal of integrated organic farm landscapes more attainable.

A special issue of the journal *Agriculture, Ecosystems, and Environment* (vol. 77, nos. 1–2, 2000) provided an overview of research accomplished through the European Union's concerted action, entitled "The Landscape and Nature Production Capacity of Organic-Sustainable Types of Agriculture," which was conducted between 1993 and 1997 (Stobbelaar and van Mansvelt 2000). This work involved a multidisciplinary team of landscape experts who visited rural areas, evaluated ecological and social factors, talked to farmers, and sought to obtain an overview of various regions. They developed a checklist that can now be used to evaluate the sustainability of any rural area; this is a comprehensive tool that combines top-down universal criteria and bottom-up tailoring to local conditions. Specifically, their six criteria for sustainable rural landscape management include environment (resource conditions), ecology (biological relationships), economy (flow of finances and services), sociology (participation procedures), psychology (subjective landscape appreciation), and cultural geography (objective regional landscape identity) (4). For each criterion, parameters were set on how to gather and evaluate data on each topic. Within the checklist framework, specific goals may be targeted for any given rural area. This provides a useful tool for studying and evaluating any rural region, and they suggest that their checklist can be used as a starting point for discussions among stakeholders, then a full study follow-up can provide specific recommendations for how farmers and rural residents can achieve landscape improvements.

Several studies employed this checklist approach in various regions. For example, in West Friesland, the Netherlands, four conventional and four organic farms were compared in terms of visual cultural geography (Hendriks et al. 2000). Major differences existed between the farms in terms of vertical coherence (soil, flora, land use), horizontal coherence (farm, farmyard, regional land use), and seasonal and historical integrity, with organic farms performing better than conventional. Organic farms offered more variety and variation, yet were visually coherent on the landscape.

The landscape checklist approach was also used to study cultural landscape appreciation and identity between organic and conventional farms in nine EU countries using evaluations by nonexperts and experts (Kuiper 2000). The nonexperts were more positive about the organic farms with feelings of "naturalness" and comfort, more sensory qualities, and historical attributes; they also feel inspired to be involved. On the other hand, experts noted that while the organic farms succeeded in improving ecological quality at the farm level, these small farms were highly diverse and did not improve aesthetic qualities of "unity" that score high for regional landscape quality. Some suggest that organic standards should require landscape guidelines in the future. This may be possible in the European context but unlikely in the United States, where rural regional planning is generally unheard of.

In the fjord region of western Norway, the landscape checklist method was used as well, this time to evaluate the contribution of two organic farms to the cultural landscape in this region of rural population decline (Clemetsen and van Laar 2000). Both farms contributed to the rural landscape: one was more historical and quaint, whereas the other was more modern and active. This reflects variations that exist among organic farms elsewhere. Overall, the experts tended to focus on the objective, measurable aspects of the two farms, since their training is ingrained. It may be better to obtain subjective viewpoints from nonexperts, as they would be looking through untrained lenses that may be more candid and revealing.

Applying the landscape checklist in Italy, two organic farms were compared with their surrounding nonorganically managed landscapes in terms of ecology, environment, and cultural geography (Rossi and Nota 2000). This comparison discovered that organic farms do contribute to the rural landscape with positive environmental (soil conservation) and ecological (biological pest management) activities that are not found in the surrounding landscape.

Finally the ecology, social, and cultural criteria from the checklist were used to evaluate two organic farms in Crete to compare them with the surrounding region (Stobbelaar et al. 2000). While there was variation between the two farms, both performed well compared with the nonorganic farms in the area. This study revealed, as did the above research in other EU countries (Kuiper 2000), that the organic farm with smaller scattered fields did not score as high on the aesthetic landscape criteria as the larger organic farm, but this may be an issue of personal preference. It seems that, to some extent, landscape beauty is in the eyes of the beholder. More notably, the environmental and social quality components are quite high for all organic farms compared with conventional farms in the vicinity.

Biodiversity

Hundreds of years ago, the very act of farming took land out of forests and led to more diversity in arable field plants, but modern agriculture has arguably destroyed this vegetation diversity (van Elsen 2000). Organic farming has the opportunity to bring back diverse, even endangered, field plants (what modern agriculture calls "weeds"). Recent research on this topic confirms that "organic cropping promotes weed species diversity" (Hyvönen et al. 2003, 131). This diversity, however, cannot be fully accomplished by separating natural and cropped areas; saving or setting aside only 5 percent of the farms for "biotopes" in which nature can develop is not true diversity. Rather, organic farms should strive to integrate biodiversity into their farming methods as a whole to create a "web of biotopes" (van Elsen 2000, 108). Before we totally discredit this argument with the comment "Who wants weeds anyway?" remember that these diverse field plants can encourage the return of more varied insects, which would provide more predators for crop pests.

Biodiversity includes all species, and one concern has been the declining numbers of birds within agricultural landscapes. Bird surveys on four types of neighboring land uses were conducted: conventional, minimum-tillage, organic, and wild land (Shutler et al. 2000). Not surprisingly, wild lands are superior to any agricultural landscapes in terms of bird habitat, but within the agricultural lands, the presence of wetlands was most beneficial to bird occurrence, and organic farms had more such lands. Otherwise, minimum tillage was found to provide cover that increased bird presence. Organic farms had more hedgerows that act as corridors and habitat for bird species. Overall, it is difficult to draw conclusions by focusing solely on separate farm fields; rather, the entire landscape must be studied to understand the effects of farms and the surrounding land uses.

Agricultural fields and field edges in England and Wales were studied over three breeding seasons for birds (Chamberlain et al. 1999). Eight of 18 species were at a significantly higher density on organic field boundaries than on conventional fields. Organic farms tended to provide a landscape more conducive to bird mating and habitat: more trees, smaller field size, and wider hedges. Even when hedgerows are present on various types of farms, researchers have found that higher vegetative species diversity is associated with hedgerows on organic farms (Aude et al. 2003). Apparently, pesticide applications on conventional farms affect their nearby hedgerows and kill some of the vegetation there as well. But hedgerows on organic farms have more diverse vegetation, which provides more habitat to attract more varieties of animal life.

Rather than looking at the birds themselves, one study investigated the abundance of centipedes as an indicator of bird diets (Blackburn and Arthur 2001). The authors believe that by studying the food supply, we can better understand the decline of birds in agricultural areas. Samples were gathered from crop field edges and adjoining woodlands from twelve pairs of organic and conventional farms in England and Wales. Species richness and diversity were actually similar between the two types of farms, but the overall density of centipedes was much higher within the organic field margins. This indicates that there was a modest overall reduction in centipedes on conventional farm fields, which would not be obvious in smaller sample studies. This decline probably was not apparent to farmers or other rural residents either, since the overall reduction was moderate and uniform among all types of centipedes (or bugs in general). Pesticides acted to diminish the population of bugs, but did not wipe them out completely – that would, in fact, be more obvious to people in the area. Keep in mind that this study reflects the somewhat smaller average farm size and presence of wooded areas in European rural landscapes, which may not be present in some agricultural regions of the United States.

Another study investigated bees on farms (Kremen et al. 2002). The importance of bees cannot be underestimated, as they provide critical pollination services and farmers usually import honey bees for this purpose, particularly as bee populations have decreased due to pesticides and habitat loss. These researchers compared conventional and organic farms located both near and far from natural habitat and found that on organic farms located near native habitat, "native bee communities could provide full pollination services even for a crop with heavy pollination requirements," in this case, watermelon (16812). Other farms sampled did not have enough native bees for pollination. This shows that well-managed organic farms can promote biodiversity, which is good for nature and for agriculture.

The International Federation of Organic Agriculture Movements (IFOAM) notes that organic farming is just a first step in the process of improving biodiversity. In fact, more research should be done to illuminate how biodiversity can be integrated into agricultural production. Another paper from IFOAM suggests that organic certification standards should be written to include preservation of biodiversity, and such conservation efforts could be regularly monitored with the usual annual on-farm inspections (Stolton and Geier 2002). Many organic farming practices promote biodiversity, including whole farm planning, omission of synthetic chemical pesticides or fertilizers, emphasis on soil health, and diversified farming systems. These activities all promote diversity and species abundance by providing increased food sources and more varied habitats.

Sustainability

Agricultural sustainability is a tricky subject, as it raises many questions: What is the time frame? What are the variables to sustain? And what does "sustainable" really mean? Scientists grapple with these questions in the context of comparing various farming methods. Research in Italy investigated the environmental and financial sustainability of conventional, integrated, and organic farming systems (Pacini et al. 2003). Although the sample was small, with only three comparative case studies, the data collection and modeling could be useful as a springboard for other researchers. An environmental accounting information system (EAIS) was used to consider "all the ecological and production processes that potentially affect the state of the agro-ecosystem" (276). Climatic and soil properties were included in the analysis of nutrient and erosion indicators (studied through groundwater loading effects), pesticide indicators (measured as risk potential), and biodiversity indicators (combining on-farm wooded areas and hedgerows with crop diversity). Financial indicators were "gross margins" that included revenues from production, government payments, costs of fertilizers and pesticides, and cost of maintenance for ecological activities. Overall, the organic farming systems performed better than the integrated or conventional systems "with respect to nitrogen losses, pesticide risk, herbaceous plant biodiversity, and most of the other environmental indicators" (273). And the gross margins were 6–8 percent higher on certified organic farms. The authors note that the overall sustainability of *any* form of agriculture must be questioned, particularly because of climatic and geographic variations across the landscape.

Edwards-Jones and Howells (2001) sought to determine whether organic farming was more sustainable than conventional farming on the source of crop inputs for crop protection and the environmental hazard of the chemicals used. They simply obtained lists of "approved" substances from the Soil Association (the main organic certification agency in the UK) and a random list of ten chemicals that "might typically be used in a conventional farming system" (41). But they did nothing to verify whether the "approved" substances are actually used by organic farmers and, if so, to what extent. Likewise, they do not provide information on the extent to which the conventional insecticides and fungicides (they don't mention herbicides, which are commonly quite toxic) are used in conventional production. Thus this article presented a hypothetical worst-case scenario for organic methods with no reference to the known hazards of conventional production. And the authors raised more questions than they answered. In fact, some very simple fieldwork and data collection would go a long way in developing a more realistic view of the variations in chemical use (thus toxicity) between organic and conventional farms.

For example, among newly transitioned organic farmers Bt (*Bacillus thuringiensis*, a bacterial insecticide that controls caterpillars or beetle larvae but remains nontoxic to other organisms) is sometimes needed in the first few years without synthetic chemicals. But as the transition is complete, the soil becomes more balanced and the farmer's organic farming skills are honed. Then they can often forego the use of this bio-insecticide. So, just because something is listed as allowed in organic farming, this does not mean that it is used very often. Many organic inputs are prohibitively expensive, so inputs are only a last resort. On the other hand, many conventional farmers are accustomed to spraying just "because I always spray." So the conventional view is one in which agrichemicals are a necessity in production, but the organic view is that purchased inputs are only for specific situations that may arise after all other means of control (crop rotation, bug picking or vacuuming, extra tillage, etc.) are exhausted.

Attempting to develop methods for evaluating farm sustainability, research can raise many interesting questions without coming to any conclusions (Andreoli and Tellarini 2000). European efforts to develop assessment strategies have included multiple criteria, such as farm performance, natural landscapes, environmental protection, and agricultural products. As noted previously, other studies indicate that ecology, sociology, psychology, and cultural geography must all be considered in evaluating an agricultural landscape. Issues raised in this article seem a bit obvious: farms vary by type and by the actions of the individual farm manager, and such complexities must be considered when formulating rural policies.

Perhaps a more telling indication of agricultural sustainability is found in a study that demonstrated how soil erosion could be drastically reduced and pest management accomplished without the use of pesticides (Pimentel et al. 1989b). Adaptive management that optimizes the biological and chemical process within farms' agroecosystems was employed and high corn yields were maintained while fossil fuel inputs were reduced by 50 percent. These are significant results in terms of agricultural sustainability, and they imply that future research into the application of organic methods is warranted.

While these farming landscape comparisons provide valuable information on the crops and methods used in organic farming, we must also ask: who is actually using these methods? Studying the farmers themselves allows us to investigate various factors that influence the decision to adopt organic methods, and we can identify the major barriers along their path. Furthermore, we can obtain a more balanced and realistic view of the entire agricultural system if we understand the individuals involved in these important daily land management actions.

FARMERS

Research methods are a key concern when seeking to describe and compare farmers. How should information be gathered? A favorite social science method is to mail out survey questionnaires and hope that farmers will fill out the forms and send them back. These surveys usually contain specific questions with little boxes to check "yes" or "no" or require that farmers select only one response: A, B, C, or D. While this is neat and tidy for data collection and analysis, we may be missing a great deal of relevant information with this approach. This goes back to the broader debate: what is "objective" research? Some people argue that we must use these little checked boxes so as to treat each farmer uniformly and to request the exact same information from each. The problem is that we may miss a great deal of relevant information this way. One farmer may check yes because she fully agrees with the question, while another farmer is confused and feels the question doesn't really fit what is occurring on his farm, but that a yes is still more appropriate than a no (even though there are a lot of unexplained issues behind that yes checkmark). In this case, a researcher obtains two yes checkmarks but under completely different circumstances that are never addressed in the study.

So, then, what is an objective survey? If we allow the farmer to "fill-in-the-blank" for open-ended questions, does that provide more realistic responses while also keeping uniformity by asking all respondents the same questions? Yes, perhaps. Yet we must wonder – are we really asking the right questions? The questionnaire may be way off the mark and skip an idea that is fundamentally important to a farmer, simply because we didn't know enough to include it among our questions. Maybe the remedy for this is to read all the past studies and base the questions on information gleaned from previous research. Ah, but what if there is something new that was not considered in other research but that truly influences the farmers you are now studying? So this debate continues. You probably know which side I am on. I think we miss a lot of important information when we strive for "objectivity." Instead we should talk to farmers and see what we can learn in open-ended conversations about their farms. This is a pragmatic approach that strives to understand the world through people's actual experience.

Comparing Organic and Conventional Farmers

Many researchers who want to study organic farmers actually undertake studies that compare organic to conventional farmers. This is a bit silly, really, as if one would say, "Hey, I want to learn about apples, so I'm going to study oranges *and* apples!" But the comparative nature of our work is due to the fact that there has been so much previous research on conventional farming that it serves as a baseline for comparison: Hey, let's see how different organic farmers are from the "norm." And this is useful in providing a context for organic farming and to illustrate what is unique about organic farmers and their field methods. And, yes, there are substantial differences between the two groups of farmers.

A study by two well-respected researchers exemplifies some of the problems with comparative research on organic and conventional farmers. This study downplays differences by asking questions that push farmers to select one response. Specifically, this survey was based on an eighteen-page questionnaire completed by 70 organic, 131 small-scale conventional, and 178 commercial-conventional farmers in New York state (Buttel and Gillespie 1988). The questions focused on crop production practices and agricultural-environmental orientations. For example, one question asked farmers their preference in crop varieties. On the questionnaire, they had three choices:

1. A variety with very high yield potential, but which requires heavy use of fertilizers and pesticides to get high yields.
2. A variety with a moderate yield potential, but which is resistant to pests and diseases so that chemicals are seldom needed.
3. Have no preference.

Wouldn't any farmer want pest-resistant varieties that reduce the cost of having to pay for "heavy use" of inputs (option #2)? And, not surprisingly, 97 percent of organic, 87 percent of small-scale conventional, and 83 percent of commercial conventional farmers selected #2. Such a structured questionnaire often masks the real motivations behind choosing various farming methods. Luckily, these researchers also included a more telling question, related to pesticides, which showed that 100 percent of organic but only 47 percent of small and 42 percent of large conventional farmers prefer natural insect control. The environmental orientations were not surprising, as organic farmers have "strikingly more pro-environmental attitudes than either small or commercial-scale conventional farm operators" (15). The idea of "preference" adds another complexity to the questions. Many people may *prefer* one thing but in fact *do* quite the opposite for various reasons. So questionnaires that force farmers to select one response from a rigid list must be clearly written and carefully interpreted.

An important scale was developed to identify people's worldview or paradigm regarding agriculture (Beus and Dunlap 1990). People who agree with the large-scale industrial modes of production operate within a conventional paradigm, while those seeking ecological and sustainable agriculture follow an alternative paradigm. The Alternative-Conventional Agriculture Paradigm (ACAP) scale was built around several issues: centralization versus decentralization, dependence versus independence, competition versus community, environmental domination versus harmony with nature, specialization versus diversity, and exploitation versus restraint. The ACAP scale, based on twenty-four questions, has been applied in several contexts around the country. Once it was applied to 208 farmers in Washington state to investigate how their agricultural perspectives influenced their actual farming practices (Beus and Dunlap 1994). There was a clear link between farmers' worldview (as measured by the ACAP scale) and their farming methods. Farmers with integrated views of conservation, environmental protection, and productivity were more likely to adopt organic agriculture.

A study of organic and conventional farmers in Michigan involved interviewing twenty-five farmers (McCann et al. 1997). This research showed that organic farmers use more sustainable methods (e.g., green manure, cover crops, hedgerows, crop rotation) and have greater concern for their farm's long-term sustainability, even if this means short-term risk. They were also more aware of ecological problems in agriculture. In fact, the organic farmers believed that "improving soil quality" was the main measure of success in farming, ranking this above "profitability of farm."

A study comparing organic and conventional growers in Massachusetts attempted to understand these growers within a rapidly urbanizing countryside (Lockeretz 1995). Thirteen fruit and vegetable growers from each farm type were interviewed, and it was found that organic farms tended to be smaller, contributed less to family income, and the growers were relatively new to farming. The organic farmers were strongly committed to organic methods and did not consider their operational problems to be overwhelming, unlike the so-called treadmill that weakens most conventional farmers. Thus organic farms could play an important role in an urbanized landscape. These highly motivated, nontraditional growers from various backgrounds may be more accepted in populated areas, especially because they do not spray agrichemicals.

A comparative study of Canadian organic and conventional farmers was based on personal interviews with 118 organic and 85 conventional farmers located in British Columbia, Ontario, and Saskatchewan (Egri 1999). In terms of sociodemographic variations, organic farmers tended to have fewer years of farming experience, higher education levels, higher numbers of women operators, somewhat smaller farms, and less hired labor. Organic farmers had high levels of environmental concern and held positive perceptions of the economic and production benefits of organic farming. Major differences were seen in organic farmers' information sources, with less than half using any governmental information sources, whereas conventional farmers commonly use government information. Organic farmers ranked other organic farmers as their most relevant, understandable, and trustworthy sources of information.

A mail survey of 156 organic and 204 conventional farmers in Australia was used to investigate work satisfaction and variations in sustainability goals between the two groups. Organic farmers were much more satisfied and found their work more interesting and challenging than did the conventional farmers. In addition, organic farmers were more likely to say that farm life was good for them and their families. They also felt that they had control over how they grew their crops and livestock and that they were working with nature. They recognized that agriculture could cause ecological problems, and they strived to use natural fertilizers and to stay in harmony with the environment. Overall, these organic farmers were highly satisfied with their work and felt confident in their knowledge and environmental competence (Rickson et al. 1999).

Interviews were conducted with both conventional and organic farmers in New Zealand to investigate how they chose their mode of production (Fairweather 1999). He developed a "decision tree" or list of yes/no questions to sort farmers into the conventional versus organic category. He found that there were a variety of reasons why farmers become interested in organic methods, such as an organic philosophy or concern about agrichemicals. Others were mostly interested in the higher prices they could earn with organic price premiums. Some were motivated by issues within conventional agriculture: perhaps they had problems when farming conventionally or they were concerned about soil quality. Others were happy with conventional farming and didn't see the need for organic methods and thus did not "go organic."

Organic farming in Austria was studied by surveying 383 conventional farmers about what they perceive as barriers to trying organic methods (Schneeberger et al. 2002). Here, conventional farmers believed the risk was too great; the perceived production challenges (weeds and pests, additional labor, decreased yields) and perceived drop in income were the main barriers to transitioning to organic. In addition, these farmers noted that they would be dependent on government subsidies, which they wished to avoid. Of course, this is particular to the European context, because no such subsidies exist to encourage American organic production!

In a comparative study of twenty-seven organic and thirty-one conventional farmers in Colorado, I focused on behavioral characteristics that were distinct between the two groups (Duram 1997). Using a mail survey and in-depth personal interviews, I defined a range of behavior from composite proactive to reactive farmers, which can be displayed as a spectrum of agroecological behavior. Thus proactive farmers, who tended to be organic producers, were more likely to find their own markets, have on-farm diversity, accept risk, trust personal experience, know ecology terms, harmonize with nature, seek new methods, feel in control of their farm, and be active in organizations. On the other hand, reactive farmers, who tended to be conventional producers, felt manipulated and trapped in the agricultural system yet still demanded its security, trusted in science and technology, specialized in a few crops, claimed to be a master over nature, but did not recognize many ecological terms. There was a strong internal locus of control among proactive farmers, meaning that they felt they had control of their farming operation and their lives, with reactive farmers feeling the opposite. Organic farmers tend to face problems head on and negotiate around barriers, often highly motivated to operate outside the status quo system of industrial agriculture (Duram 1997).

Overall, then, we see that comparative research on organic and conventional farmers shows that organic farmers tend to have higher education levels, have less previous on-farm experience, complain about a lack of information on organic methods, and have more environmentally friendly attitudes. Also, more women are active in organic farming than in conventional production. There are also some regional variations and differences among farm types (large and small-scale issues). Clearly, variations exist between farmers who have chosen organic versus conventional methods.

Focus on Organic Farmers

A few studies focus solely on organic farmers, describing them on their own terms without taking a comparative organic to conventional approach. In an early study, the research question was simply "Who is the organic farmer?" (Dalecki and Bealer 1984). This survey of eighty-seven organic growers across the United States was spurred by the authors' disagreement with an earlier study of Michigan farmers that found the average organic net farm income was $0, thus "proving" that organic management strategies were not profitable (Harris et al. 1980). In fact, the second study found that organic farmers had high levels of education, earned a moderate amount on their farm, and marketed to retail stores, farmers' markets, roadside stands, and cooperatives. This contradicted the earlier study particularly in terms of education levels and farm income.

In a 1983 study, fifty-eight organic producers and thirty-two consumer members of a Kansas organic organization were surveyed (Foster and Miley 1983). This early study already depicts the lack of information sources available to organic farmers – 95 percent of the farmers would not contact extension agents, USDA, or university researchers for help with farm practices. But overall this exploratory research identified the link between organic producers and consumers, as both groups think food production is a social concern and believe that they should support their local communities by selling/buying food locally. An update of this type of study would be particularly relevant today, as organic consumers are a rapidly growing segment of food shoppers.

In our attempt to gather more complex survey information, rather than forcing surveyed farmers into narrow simple categories through confining questionnaires, we run the risk of opening a can of worms. This is healthy for soil, but not for social research! Organic farmers are a complex and varied group, so it can be difficult to identify the personal characteristics that influence the structure of their farms and their perceptions of organic agriculture (Lockeretz 1997). Forty-three organic growers in Massachusetts and Vermont were interviewed, and it was found that the growers' personal characteristics varied considerably and did not relate to how their operations followed organic principles or how they perceived problems in organic agriculture. At least in terms of trying to find statistical significance, it is "a great oversimplification to lump all organic growers into a single group," and we must go beyond seeking only a handful of "personal variables" to a more intensive approach (Lockeretz 1997, 23).

Focusing on organic farmers in Illinois, I sought to understand the obstacles and opportunities faced by organic farmers (Duram 2000). The only way to fully describe their reality was for them to tell the story themselves, so I conducted twenty in-depth interviews with farmers across the state. This yielded 435 pages of transcribed interviews that were fascinating. I read these over many times to try and pull out the themes – what were these farmers, as a group, really saying? And I developed a list of factors that seemed to influence Illinois organic farmers. Some of this work led to larger questions of regional variation and influences, which I am now able to address in this book.

By interviewing thirty-five farmers who converted to organic methods and then reverted away from organic, researchers discovered key factors that motivated the initial attempt at organic methods and also the reasons for quitting (Rigby et al. 2001). There seemed to be two main paths for this adoption/reversion process: first, farmers motivated mostly by economic gains reverted because they could not realize adequate sales or prices to offset the higher costs of organic production; second, farmers who were motivated by lifestyle or ideals but had little experience or knowledge tended to quit organic farming because they could not make a living. Overall, the main reasons why farmers ceased organic production were marketing and market incentives, cost issues, agronomic problems, and lack of technical assistance.

Kaltoft (1999) investigated how knowledge, views of nation, and farming practices affected Danish organic farmers. By focusing on six in-depth interviews, Kaltoft identified different worldviews that describe the variations among the farmers who focus more on soil, nutrients, or biodynamic preparations. It is interesting to note that Denmark established national regulations in 1988, nearly fifteen years before the United States, and has actively encouraged farmers to adopt organic methods. This "institutionalization" of organic farming influenced farmers by making it more socially and economically acceptable, but this in turn may diminish the original philosophy behind organic farming.

Grossman (2003) investigated knowledge of soil processes among organic coffee producers in the Chiapas state of Mexico. Grossman interviewed thirty-one members of a certified organic coffee-producing community that grew shaded coffee. This is environmentally beneficial because it maintains forested areas. In summary, the farmers understood what is beneficial to build the soil (incorporation of leaves and other compost, planting legume cover crops, earthworm populations) but not necessarily the physical processes of why it works (soil mineralization, microorganisms in decomposition, nitrogen fixation from legumes). This study exemplifies the care researchers must take in investigating ecological topics. The farmers clearly understood the intricacies of growing organic coffee, but were not familiar with the science or the scientific terminology behind what they saw in the fields.

Research into the specific pest management activities of organic farmers was conducted using the Organic Farming Research Foundation's 1997 national survey of 1,192 organic farmers (see Walz 1999). A statistical model was built to show the types of farmers most likely to experiment and try various pest management techniques (Lohr and Park 2002). Findings suggest that "college-educated farmers with smaller acreages, more than half their acreage in horticultural production, and extensive experience with organic production methods, have the greatest diversity in their insect management portfolios" (87). If farmers had access to reliable information, they would employ more alternative management techniques. It seems that public research should step up and fill this information gap.

The "diffusion of innovation model" is a generally accepted concept within agricultural technology, in which there are a few innovators who first try something new, then a majority of adopters take it on, and finally a few laggards adopt it quite late. Research investigated whether organic farmers fit the traditional "diffusion of innovation" model, and it turns out that organic farming is a complicated process that doesn't completely fit here.

Instead, organic methods are developed from a self-motivated approach with just a few pioneers, which is quite different from a new mainstream agricultural technique that is introduced broadly. So organic farming is more of a bottom-up innovation that means a very slow diffusion rate will occur. In addition, policy support and market development are critical issues that influence adoption rates (Padel 2001).

The general conclusion of research on the topic of organic farmers is that they are a diverse bunch of folks. It is hard to pinpoint social or demographic characteristics because there is just so much variation among these farmers. But in general they have strong convictions that have compelled them to do something different. They have the courage to face risk and go against the status quo of industrial agriculture. They are independent and innovative. They find information, solve pest and soil concerns on their farms, and market their numerous crops to meet growing consumer demand for organic food.

3

The Social Context of Organic Farming

Good farming is farming that makes people healthier. It does so by creating and delivering food of the highest attainable nutritional quality and safety. Agriculture fails in its most obvious mission when that quality of healthfulness is missing or when it becomes corrupted by such things as toxic residues. . . . The problem is compounded by . . . numerous processors who, for the sake of profit, have been known to take most of the nutrition out, put additives in, turn wheat into Twinkies and corn into breakfast-table candy.

– Donald Worster, "Good Farming and the Public Good" (1984)

ccording to one recent survey, 70 percent of Americans have purchased an organic food product at least once, with 32 percent buying occasionally and 16 percent buying organic every time they shop (Gardyn 2002). Sales of organic products are booming, and consumer interest in organic items shows no signs of slowing. While organic trade companies and marketing firms strive to understand shoppers' motivations, organic farmers also seek information on consumer trends. So who buys organic food? Not surprisingly, research shows that the answer is complex.

ORGANIC FOOD CONSUMERS

A review article provided data on consumer demand for organic foods based on information from twelve other research studies done between 1987 and 1997 (Thompson 1998). There was a great deal of variation in the demographic characteristics of organic food purchasers. Specifically, income, education, and age showed opposing high and low trends: lower income (<$25,000) and medium-high income (>$50,000) consumers were more likely to buy organic foods, as well as the age groups 18–29 and 40–49. Likewise education was a complicating factor: apparently college-educated consumers seemed more likely to buy organic food, but those with graduate degrees were less likely. Additionally, it is important to consider store choice as an indicator of organic purchase likelihood, since there is variation in where organic products are available. A more recent survey shows that people ages 18–24 and 35–49 are most likely to buy organic products, and 20 percent of consumers in the northeastern United States buy organic products every time they shop (Gardyn 2002). Organic purchases were motivated by health concerns and the idea that "organic food is healthier for me and my family" more than environmental concerns. Another poll indicates that over half of shoppers are more likely to buy food labeled organic, with women having particularly positive opinions about organic foods (ABCNews.com 2001).

Overall, demographically diverse people are motivated to buy organic food for various reasons. Age may influence their memories of time in the countryside, or their education may teach them about soil qualities, or their income may influence them to spend their dollars carefully, or parental concern may cause them to scrutinize their children's food, so they intentionally purchase organic food.

International Perspectives

A study conducted in Britain looked at the key reasons for growth in consumer demand for organic food in that country (Ilbery et al. 1999). First, it was promoted as environmentally friendly, and people saw it as a solution to the social and environmental problems of industrial agriculture. Second, consumers turned to organic foods because of food scares such as mad cow disease and concerns over genetically modified organisms (GMOs). Third, consumers were becoming less accepting of mass-produced food and inhumane treatment of animals in industrial production. Fourth, a lifestyle choice, perhaps related to socioeconomic status, influenced consumer purchases of organic food. Finally, organic farming was becoming more accepted within rural areas and among farmers themselves (at least in the UK), and this may remove some of the social risk previously associated with producing and consuming organic food. A related issue is that regional identity can influence consumers' decisions to purchase specialty food products made in their local area (Ilbery and Kneafsey 1999)

In many European Union countries, livestock products are among the top organic products purchased (Hermansen 2003). There is both increasing consumer demand for organic foods as well as increasing farmer interest in producing organically, which is stimulated by government support. Organic farming must "fulfill the expectation of each of these stakeholders if organic livestock production is to increase further" (3). Further, consumer interest in organic livestock is based on concerns about personal health, the environment, and animal welfare. Hermansen's article is one of several in an issue of *Livestock Production Science* (vol. 80, 2003) that focus on organic systems.

Magnuson et al. (2003) investigated consumer choices in Sweden. A survey of 1,154 Swedes showed that "self-reported purchase of organic foods was most strongly related to perceived benefit for human health" – both of the individual and the family (109). In addition, consumers with environmentally friendly behavior were more likely to purchase organic products.

A comprehensive study of both consumers and producers was undertaken in Norway (Storstad and Bjørkhaug 2003). In fact, four groups were defined and analyzed: 912 conventional consumers, 55 organic consumers, 439 organic farmers, and 383 conventional farmers. Overall, organic consumers and organic farmers had similar opinions regarding the environmental problems from agriculture, while conventional consumers have moderate views, and conventional farmers have the least criticism of agriculture's impacts. Storstad and Bjørkhaug also discuss how consumers might be motivated by broader distrust in technological fixes to social and ecological problems.

Researchers in Greece used certain interviewing techniques to construct "means-end chains" that describe consumers' behavior (Fotopoulos et al. 2003). Their focus was consumers' reasons for purchasing wine produced from organic grapes, but their research approach could be used in many contexts. They found deeper reasoning and motivations behind organic purchases, and the issue of quality came through loud and clear. Organic shoppers were influenced by several notions in combination: quality, seal of approval, country of origin, and pleasure. So linking these concepts may benefit future marketing of organic foods.

Complex factors influence people's motivations for eating organic food (Lockie et al. 2002). In Australia, both focus groups (thirteen groups of approximately ten people each) and a broad telephone survey of twelve hundred people were used to investigate whether people ate organic food and the reasons for their choices. First, the focus groups yielded more strongly opinionated responses, which is often the case since more vocal people tend to dominate focus groups and often sway the discussion. These groups offered more negative views of organics, from higher cost to distrust of organic methods. They also expressed the common stereotypes of organic shoppers as hippies and health nuts. From the telephone survey, however, we learn that over 40 percent had eaten some organic food within the past year. More women and more educated people tended to consume organic food. People in all income levels ate organic food, with wealthier people able to purchase it more often. Organic consumers were concerned about price, but balanced this against a range of other factors, such as health, natural ingredients, animal welfare, environmental protection, fitness, and political values. These consumers were more likely to perceive problems in industrial agricultural methods, especially biotechnology.

U.S. Regional Variations

Due to the geographical complexities of attempting to study organic shoppers at the national level, organic consumer trends have mostly been studied at the regional level in our country. Beginning in the early 1990s, survey research focused on trying to identify the various groups of organic consumers. In upstate New York, 350 shoppers were contacted at a food cooperative and surveyed with written questionnaires (Goldman and Clancy 1991). This group was predominantly female (62 percent), averaged thirty-five years of age, had some college education (93 percent), had incomes under $20,000 (51 percent), were vegetarian (34 percent), and purchased most of their groceries at the co-op (49 percent). Over 40 percent of the shoppers almost always purchased organic produce and were minimally concerned about insects and surface blemishes. They were more concerned about pesticide residues, and they were less price sensitive than other shoppers. One-third said they were somewhat or very likely to pay up to 100 percent more for organic produce. Despite their relatively low incomes, many purchased organic products because price was less important than nutritional value.

Likewise, mail surveys of 389 members of a Georgia consumer panel for food research found that 61 percent preferred organically grown produce (Misra et al. 1991). Variables such as age, race, education, and household income were not clear predictors of preference. Factors such as freshness, appearance, and nutritional value influenced their produce purchases. Consumers with concern for pesticide residues, preservatives, and nutritional value were more likely to tolerate some blemishes on the organically grown produce. Consumers were willing to pay 10 percent more for organic produce (66 percent said yes), but not so willing to pay more than that, with 22 percent saying they would pay no price premium.

Information from 1,769 mail surveys was used to assess California consumers' perceptions of conventional and organic food to understand who buys organic food (Jolly et al. 1989). Regardless of their age, income, and gender, 23 percent indicated that they look for organic food when they shop and 30 percent said they plan to buy organic food within the next month or so. And 57 percent of consumers ranked organic food as better than conventional food on the basis of food safety, freshness, general health benefits, nutritional value, effect on the environment, flavor, and general appearance of the product. These consumers were willing to pay 30 cents more per pound to buy organic carrots, broccoli, apples, and peaches.

RETAIL SALES AND SUPPLY

Motivation for buying organic food is linked to information and labeling. A study was conducted on the topic of promoting sustainably produced apples in Minnesota (Robinson et al. 2002). In this case, the food was not organic but had a seal of approval from the Midwest Food Alliance. Customers were surveyed initially and again after an eight-week educational and public relations campaign. Clearly the campaign had an effect on the consumers' knowledge, thus we can assume that other educational and in-store marketing interventions would help inform consumers about organic foods as well.

Focusing on retailers, a survey of California supermarket chains investigated the perceptions of produce and marketing managers toward organic produce (Jolly and Norris 1991). This study was framed within the context of the food scares of the late 1980s (Alar on apples and cyanide on Chilean grapes). These events sent consumers scrambling to buy organic produce, but as demand increased suddenly and rapidly, supply fell behind and prices soared. This actually led to a negative situation for organic food, as many stores stopped stocking organic foods because the high costs made them impractical. This study shows that by 1991, many store managers were interested in organic produce; it rated better on environmental impact and residues but worse on appearance and shelf life. Overall these stores wanted to carry more organic products if supplies were available.

Other researchers sought to inform retailers about "consumer perceptions of pesticide residue concerns" (Byrne et al. 1994, 492). But the authors neither provided a definition of certified organic food nor investigated consumer motivations for it. They tried to predict whether people would likely shop at a supermarket carrying this mysterious "pesticide residue free" produce (again, not using the words *certified organic*). Their complex statistical models were not based on a specific dollar amount – just the wording *higher priced*. They claimed that "this costlier commodity could negatively impact the consumer's price image of the store" (499) and "this result suggests that more than one in five consumers would not shop at a supermarket offering this higher priced produce" (500). This fuzzy research did not define the important terms: *certified organic* and *higher priced*. Unfortunately, the findings were targeted at retailers, which could make them wary of stocking organic products.

Price and production quantity data for organic broccoli, carrots, celery, romaine lettuce, strawberries, and watermelons were used to model "supply elasticities" – or how price influences the amount of crops grown (Lohr and Park 1995). For the 1985–89 period, wholesaler prices for these organic crops, production quantity, and climate influences were statistically modeled. They found that organic farmers have more marketing flexibility because they can, and do, sell through either conventional or organic outlets. Their model indicated that higher prices would lead to increased supply of organic produce, with carrot output quicker to adjust and lettuce slower. As the organic produce market matures and develops a "structure of the industry" (29), these supply issues will vary. The question is, do we now have a mature organic market? How does this affect consumers and farmers? Locally the answer is not in doubt, as farmers have feedback on consumer demand, in the form of personal contact, and can vary crops as needed. In larger distribution channels, farmers still tailor their crops for certain markets, but once the crop is in the ground there may be a long lag time before their specific crop rotation can include major changes in crop types.

RURAL GEOGRAPHY AND LOCAL FOOD

A food mile is a geographic concept used to indicate the distance that food travels from farmers to consumers. Using produce arrival data from the USDA Agricultural Marketing Service, the Chicago terminal market was examined for 1981 and 1998 (Pirog et al. 2001). A weighted average source distance was used to calculate a distance estimate that combined information on the distances from producers to consumers and the amount of food product transported. Produce arriving by truck traveled an average distance of 1,518 miles to reach Chicago in 1998, a 22 percent increase over the 1,245 miles traveled in 1981.

The current global food system is geographically illogical. For example, it is supposedly "cheaper" to sell citrus from Latin America in northern

Europe, while citrus from nearby Mediterranean countries is dumped (La Trobe and Acott 2000). But, of course, the full costs of transportation are not included in the price. If true environmental costs were figured in, production and consumption would occur in closer proximity. Using food miles as the basis of our argument, we can see potential benefits with organic production. Notably organic farmers should seek to distribute their crops locally and regionally through direct marketing. In this way, organic agriculture has the potential to take out the "middleman." This keeps consumer prices down and provides more profit for the farmers who grow the food.

Direct Marketing

Farmers' markets are one way to market organic crops locally. These are a traditional means of bringing together community members and linking local farmers to nearby consumers. Farmers' markets now range from informal, once-a-week gatherings of ten farmers selling out of the back of their pickup trucks to highly organized regional markets with a county fair atmosphere of crafts, music, and food (which may be locally grown). Some markets do not have a rule that a vendor must produce the crops they sell, so ask at your market. There is now a resurgence across the United States in the number and size of farmers' markets (Brown 2002). According to the USDA, the number of farmers' markets in the United States has increased 79 percent between 1994 and 2002. The 2002 USDA National Farmers Market Directory shows there are over 3,100 farmers' markets in operation, and undoubtedly some smaller markets missed the tally (USDA–AMS 2002). Clearly farmers' markets constitute an important direct marketing tool for many farmers, whether as the sole outlet or as just one component in their total marketing activities.

Research on farmers' markets has investigated characteristics of the vendors (Lyson et al. 1995). In the state of New York, these researchers selected nine markets and obtained information from 115 vendors who were identified as either full-time growers, part-time growers, or craftspeople. Across all groups the two most important reasons for selling at a farmers' market were not economic. The vendors simply enjoyed it and liked talking with customers. Next came the economic reasons, of course: the vendors need the income. While nearly all the full-time growers lived in rural areas, fewer than half of the craftspeople did. Full-time growers noted that only 35 percent of their gross sales are from the market, with the remainder mostly from other direct market methods (30 percent) and wholesaling (19 percent).

Part-time growers sold mostly to farmers' markets, with 68 percent of their gross sale here. Interestingly, 21 percent of full-time growers were female, whereas conventional agricultural operations are mostly led by males. And 37 percent of full-time growers had hired labor, indicating that operations were of significant size to warrant this help. Clearly this work substantiated the idea that farmers' markets address both an economic and social function for vendors. These authors noted that much economic theory overlooks markets because it ignores these community relationships which emerge within the local economy.

Community Supported Agriculture (CSA) is another means of direct marketing that is rapidly increasing in popularity. Some people actually call it Community Shared Agriculture (Fieldhouse 1996). It provides a means for linking farmers and consumers and developing a true sense of community (Norberg-Hodge 1995). CSA is a specific connection between a farm and a group of customers who are paying "members" or "subscribers." This relationship allows for the members to share in the production risk (e.g., will there be a drought that destroys all the beans or a flood that washes out the lettuce?) rather than the typical situation in which all risk is assumed by the farmer. Likewise, the members share in the crops harvested. Usually, a fee is collected upfront before the growing season, when a family or individual joins the CSA. Then for a certain number of weeks during the spring, summer, and fall, the member receives a basket or bag full of fresh, in-season produce. Each CSA has its own "culture." Some require members to work a few days during the growing season – planting, weeding, picking, or boxing the produce. Others do not require on-farm work, but encourage farm visits through field days and other events. Distribution of the weekly produce baskets varies by group as well, with some CSA farmers delivering these to a central drop-off point in the city, and others having members drive out to pick up the produce. In either case, these CSAs provide strong links between rural production and urban consumers, and most are located within easy driving distance of urban or suburban areas.

The informative guide *Sharing the Harvest* described the background and development of CSAs, as well as specific concerns such as legal aspects, labor, and food distribution (Henderson and Van En 1999). Topics were addressed through examples of successful CSA operations. For example, a CSA may opt to be less "efficient" but more educational and participatory. Likewise, social issues such as sliding scales for fees and accepting food stamps were explored as ways of attracting members from all walks of life.

Research on CSAs investigated both farm characteristics and member motivations. Near Minneapolis, eight CSAS were studied via observation, interviews, harvest survey, focus groups, and a survey (Cone and Myhre 2000). These CSA farms averaged 6–10 acres of vegetables and varied from 30 to over 300 members, operated by farmers who were college-educated married couples, with little previous farm management experience. Key goals as noted by these farmers were sustainable production and nurturing the environment and community. The consumer members were mostly educated, affluent urbanites who said that their reasons for participating in the CSA were environmental concerns and the ability to get organic food that is fresh and local. The researchers wisely interviewed the CSA "dropouts" as well, who noted that they quit because of the inconvenience and lack of variety in the weekly produce baskets. Overall, members with higher participation in the CSA, as seen in attending events and visiting the farm, were more likely to be motivated by land stewardship, supporting a local farmer, and belonging to a community. Yet we must question how people interpret and define a community. These authors note that CSAS are a "growing social movement that endeavors to make direct connections" between farmers and consumers (Cone and Myhre 2000, 187). But only half of the members they studied actually participated beyond buying and eating their shares.

One university researcher obtained grant money and began a CSA with the hope of building a community (DeLind 1998). The reality turned out quite different from her group's initial community "missionary zeal" (5). The farm grew too quickly and was soon driven by an economic mind-set rather than a community mind-set. For example, members would take given amounts of produce even if it was too much for them, just to "get their money's worth." So the CSA became just another market rather than a true alternative to a market economy. Most notable was the majority of members' lack of sharing in the work or even being appreciative of the hard work of the few. She notes that members coming to pick up their produce would stand in the garden chatting without ever bending over to pick a weed. In some way, maybe this is "community" for them; everybody has different expectations.

Should CSAS have the deeper responsibility to change the social relations within agriculture, as postulated by many researchers? To put the colossal task of social change on the overworked backs of CSA farmers seems an unfair burden. CSAS could be just an advantageous way to direct market organic produce, a means of selling and buying that benefits the farmer (with higher profit margin) and consumers (with fresh local food) by taking out the middleman. And CSAS could also be an initial way to educate busy urbanites about rural issues and organic farming.

Local Organic Food Systems

Geographically, the idea of a "foodshed" is intriguing (Kloppenburg et al. 1996). Like a watershed, within which all water flows to a common stream, a foodshed is an integrated local system within which the activities of farmers and consumers flow in unison. That is, food miles are reduced and profits stay within the local area; environmental stewardship is everyone's concern and consumers are willing to pay the true cost of food production.

Local food systems may be an "appropriate way to revitalize a community" (Feenstra 1997, 28) and can be initiated by consumer advocacy. There are many grassroots efforts currently in place that promote local food and communities (Henderson 2000). Food production and markets are based on local environment and health goals. Key issues include learning how to eat seasonally (in most places you cannot expect fresh peaches in January!) and assessing what consumers know about and demand from local food. Many creative community food system projects are currently in action across the country. Several steps are important in the initiation of such a project. First, learn about your local food system: gather historical information on agriculture in the region, estimate your area's food self-reliance potential, and identify local seasonal food sources. All these activities must be tailored to a given geographical location and should include public participation, new partnerships, and a commitment to social, economic, and environmental justice concepts (Feenstra 2002). A wide array of people and groups – agricultural, environmental, religious, and charity groups could provide a promising network. Success is more likely to be achieved with broad participation.

Research linking rural development and organic farming suggests that there are inherent values within the organic movement that could act to encourage local food networks and local community involvement. Pugliese (2001) notes that organic agriculture and sustainable development in rural areas can be linked through innovation, conservation, participation, and integration. These four stepping-stones are necessary for rural development, and organic farming can successfully combine them within an appropriate social context. Innovation is related no longer to technology but to an attitude that combines creativity and risk. Conservation of rural resources can be achieved by combining modern techniques within a traditional context of organic farms. Participation of both producers and consumers occurs when farmers have a pro-active attitude and pride in their farms. Yet integration of social and ecological factors in rural areas can be promoted through organic farming that acts to reorganize agriculture and diversify rural economies.

Pugliese notes that within the European context, there are obvious synergies between organic farming and processes of rural development, which can be encouraged with political support in the future.

A study in Norway looked at the overlap between consumer and organic farmer perceptions of their rural region (Torjusen et al. 2001). Research methods included a "vision seminar" to discover the status of agriculture in the region, a rapid appraisal approach that involved farmer interviews, and a mail questionnaire of 368 consumers (54 percent of whom had purchased organic food). Many interests of organic consumers and farmers coincided on topics that go beyond simple organic methods listed in certification rules. The authors note that these groups both tended to be concerned about ethical, environmental, and health issues and were locally oriented, which indicates a great potential for building a local food system in theory and in practice.

So, organic farming can help us reduce those food miles! It has the potential for aiding rural development efforts, through local food distribution and personal relationships among farmers and consumers. But we must be cautious, as there are factors that may inhibit this synergy. Specifically, organic production and marketing may become as globalized as conventional production, if current trends continue.

AGRI-BUSINESS GOES ORGANIC

There are indications that organic farming is becoming more like conventional farming, a process that has been termed "conventionalization." Perhaps the most devastating exposé of the new industrialization of organic agriculture was Michael Pollan's article (2001b) in the *New York Times Magazine*. Even the most ardent organic supporter can't help but feel disappointed and perhaps a bit betrayed by the way processed organic foods are so similar to conventional ones. Personal interviews with many organic industry experts showed how big business made inroads into the organic market. He described, for example, an organic TV dinner that contained natural chicken flavor and xanthan gum (two rather unnatural sounding ingredients) and noted that the brand (Cascadian Farms) was actually a subsidiary of the huge U.S. food conglomerate General Mills. The story of organics, as he tells it, is one of big money business, and it contradicts traditional organic philosophy.

Huge organic vegetable farms (often acres cut out of the middle of a mega-conventional farm) are found in California. According to this article, the "industrial production" of both organic and conventional crops uses similar machines and crews, and only the actual fertilizer and pest control methods are different. The organic fields rely on horse manure compost and beneficial insects rather than agrichemicals. But despite this important difference, such production practices undermine the fine line between conventional and organic production, particularly in terms of scale. Large corporations such as General Mills demand large supplies. It is simpler for them to buy from a single 2,000-acre farm than from twenty 100-acre farms.

This "conventionalization" of organic farming appears to be a regionally specific phenomenon, based on California production. We can trace this from the development of the California Earthbound Farm brand, which began in the early 1980s with a 2.5-acre organic raspberry farm. It is now a brand grown on 11,000 certified organic acres in three states (California, Colorado, and Arizona) and three countries (United States, Mexico, and New Zealand). In addition, the Earthbound Farm brand is actually part of the Natural Selection foods company, which controls 6,000 more acres under the Mission Ranches company and has a close relationship with the large conventional produce company Tanimura and Antle. Tanimura and Antle recently converted some of their huge acreage into organic production (Fass 2002). At least in terms of the California example, consumers are fooled into buying organic products that are supposed to be more "natural" when in fact they are produced with the same socially "unnatural" large-scale methods as their conventional counterparts (Guthman 1998). In terms of both consumers' and farmers' perceptions, there is variation in how "natural" organic farming truly is (Verhoog et al. 2003).

Based on interviews with approximately seventy experts in the organic sector (certifiers, farmers, processors, retailers), Buck et al. (1997) describe the "conventionalization" of California organic vegetable production, particularly in the concentration of marketing and distribution. The trend is for organic farms to "look" more like conventional farms in terms of migrant labor use, purchased inputs, and off-farm processing and packaging. As the system evolves, it could be that smaller organic farms will be overlooked, as marketing opportunities and distribution systems become tailored for larger production. Unfortunately, these authors provide no tangible suggestions for how we can slow this trend.

More specific research on organic growers' practices suggests many California farmers are only half-heartedly applying organic techniques to their crop management (Guthman 2000). This study is confusing, however, because both certified and noncertified organic growers were included, so their techniques are not uniform or inspected. In any case, 150 growers were interviewed, and their field techniques were rated according to accepted organic methods. Results were not favorable. Large-scale production and organic production on "mixed" conventional and organic farms were furthest from organic ideals in terms of integrated fertility, diversified cropping, and biological pest management. Some of these growers are just reaching the minimum acceptable standard, and of course that is a far cry from what most people consider the inherent philosophical underpinnings of organic farming: local marketing, on-farm cycling, crop diversity, and small-scale direct sales. Other researchers go so far as to claim that two groups have formed: the organic food industry (and most organic growers) versus a small group of organic movement farmers (Goodman 2000). The former is completely removed from the broader goals of the sustainable agriculture movement. Instead, they focus on the organic market share and lists of regulated materials that are allowed or prohibited in certified organic production.

But these "conventionalization" studies were all based on analyses of organic production in California; clearly regional variation exists within the United States and elsewhere. Organic farmers in other parts of the United States (and even some in California) are motivated by broad goals of sustainability, both ecological and economic. I have found organic farmers in Colorado and Illinois to be motivated by strong pro-active attitudes and the determination to operate outside the conventional agribusiness mode of agriculture (Duram 1997, 2000). The farmers you'll meet later in this text seem to defy the "conventionalization" argument, but we can see how it is a slippery slope that we must dig in our heels to avoid.

The idea of "conventionalization" was also rejected in a Canadian study of organic farmers (Hall and Mogyorody 2001). Their methods involved a telephone survey of 259 farmers, 18 case study farms, and interviews with members of key organic groups. They found that in Ontario there was little evidence of organic farming evolving into conventional. "Government and conventional agribusiness capital remain reluctant to fully embrace and promote organic field crop farming" (418). This is partially due to the strong alternative orientation of a "critical mass" of producers and consumers who have an ideological opposition to conventional agriculture. Specifically, organic farmers feel that their quality of life is of prime importance, and they value the independence and diversity of their work.

Investigating the issue of "conventionalization" in organic farming in New Zealand, research drew from case studies and found that the industrial agricultural system leaves "profitable and sizeable niches for small-scale producers" (Coombes and Campbell 1998, 141). Because of inherent biological conditions and the alternative social movements behind organic farming, agribusiness manipulation will be difficult, and small-scale organic producers will endure.

Drawing from an Irish example, one analysis indicated that organic farming has been swallowed by the conventional political structures of agriculture (Tovey 1997). By claiming to support organic farming, the large-scale "status quo" policies of conventional production become stronger in the long term. Thus in Ireland, the Rural Environmental Protection Scheme funds from the European Union are promoted for organic farmers, but this will not change the basic problems of large-scale production and overproduction. Organic farms are being used for environmental conservation, in a popular attempt to keep the countryside looking pretty. In a sense, organic farms are being used as an excuse – something that can be pointed out: "See, we're trying to change the problems with conventional agriculture." But, in fact, the opposite is true, since organic farms are being funded by the existing agricultural system for its own public relations benefit, rather than promoted as an alternative approach.

A comparison of the institutional framework of organic farming in Denmark and Belgium found that each country displays different relationships between conventional and organic agricultural organizations and within organic farming groups (Lynggaard 2001). Denmark has had "creative conflict," but eventually positive relationships developed between conventional and organic marketing institutions, which has led to vast expansion of organic farming and large customer demand. Belgium, on the other hand, has had slow development of organic farming because the institutions developed regionally and separately from conventional agricultural organizations. This implies that some mutual relationship with conventional agriculture has a positive effect on organic farming, by increasing its marketing potential. But obviously, too much of a good thing must be avoided. Organic farms should coexist with, but not be subsumed by, conventional agriculture.

There are tangible, negative aspects to the agribusiness takeover of organic agriculture, and consumers and farmers should work to halt this process. But don't write off organics just yet! We should remember that organic farming is becoming mainstream – for better and for worse. If the evolution of organic farming includes an inevitable movement toward a conventional agricultural distribution system, at least it means we are not spewing as many unnecessary agrichemicals into our environment and perhaps our waterways will be just that much cleaner. In addition, more consumers will have the option to buy organic if it is more widely distributed with reasonable prices. But the most convincing reason to slow the conquest by

Big O Ag is to protect family organic farmers; otherwise, organic farms will replicate the current industrial agricultural system, in which multinational corporations control inputs, marketing, processing, and sales. The main lesson here is to support family organic farming by buying organic products – the more local the better. The better you know the farmer who produced your food, the more you know that you are supporting organic farming. In addition, stay on top of organic regulations and voice your concerns to the USDA National Organic Program, your senators, and your representatives. Don't let Big O Ag jump in and erode organic standards for its benefit.

International Trade

An article in the on-line journal *Salon* notes the irony of using massive amounts of nonrenewable energy to ship organically grown food halfway around the world (Baker 2002). The example provided is of an American working for a German organic certification agency who flew to Japan to inspect a food processor who uses Chinese soybeans to create tofu for a European market. While this provides an interesting geography lesson and in fact sounds far-fetched, it does prove that organic products are increasingly global. The United States exports organic products mostly to northern Europe, Canada, Australia, Japan, and China (Lohr 1998). This globalization raises numerous questions about the sustainability of such global marketing and the fundamental goals of organic – traditionally, those of local production and sustainability. Keep in mind that in the United States, the national certification standards are run by the National Organic Program, which is part of the USDA's Agricultural Marketing Service, so clearly the economic goals of national and global marketing are influential.

There are complex policy and economic factors involved with importation of organic foods (Barrett et al. 2002). For example, with demand for organic products high in the UK and several other European Union nations, imports are necessary to fill consumer needs. Even as British organic production is increasing rapidly, the UK imported 70 percent of all organic food sold there. While the United States is a major source of organic imports to the European Union, it is interesting to note that nearly sixty developing countries have import authorizations as well. Organic methods hold great promise for ecological and social benefits in developing countries for *domestic* use which should be encouraged (Rasul and Thapa 2003). There are substantial barriers for farmers in developing countries to export their products – most notably cost for certification and acceptance of various organic labels from abroad. And as much as possible, we should try to encourage regional marketing of food, shorten the food miles, and keep profits locally.

One way to help local farmers and slow the agribusiness takeover of organic distribution, at least in terms of the international trade of organic products, is by linking organics to fair trade. The fair trade movement seeks to transform the social context of agriculture, to promote equitable and sustainable production and marketing relations. Consumers pay for these practices, since they are educated through specific fair trade labeling (Raynolds 2000). Thus a consumer may be willing to pay extra for free trade coffee because they know the local farmer is earning a fair price and growing in a sustainable manner. On a related topic, there may be important connections between organic production and ethical trade (Browne et al. 2000). Interviews were conducted with retailers, trade organizations, importers, and agency representatives in the UK to investigate definitions of ethical trade and the means for linking it to organic production. In the UK at least, the authors predict more overlap between the two, as consumers demand that internationally traded organic produce is grown with ethical standards. The best current example is organic free trade coffee; these mutually beneficial relationships could be accomplished through joint certification. In the future, the global trade of organic products should be linked to ethical trade ideals, combining the positive ecological and social goals of both.

AN ORGANIC SOCIAL MOVEMENT?

Having raised the question of whether organic agriculture is evolving into an agribusiness-controlled Big O Ag system of production and consumption, we should now ask the opposite: to what extent does organic farming reflect an ideal grassroots social movement? Many sociological and anthropological studies have investigated this issue. There are two main points to ponder. First, many organic farmers do not have time to debate or philosophize over these deeper complex social issues. They are busy growing and marketing crops and livestock. Second, many of these studies are highly theoretical and forbidding. But I've done my homework, and now I'll concisely present some key studies on organic farming as a social movement.

Discourse analysis was employed to find linkages between the New Zealand organic movement, export standards and certification, and global consumer demand (Campbell and Liepins 2001). The argument is that organic agriculture is "exceptional" and will not follow a linear trajectory toward "conventionalization"; rather, "the processes that form the discursive field are somewhat more circular and becoming increasingly complex" (36).

In other words, organic farming in New Zealand remains closely aligned with the organic social movement that stands in opposition to the industrial food system. The authors show how the development of Bio-Gro organic certification in the 1980s was accomplished by organic farmers, biodynamic growers, inspectors, exporters, and consumers and other stakeholders working together. So "by institutionalizing one discourse of 'organic' " (that is, the creation of organic certification standards), there was a shift in "the terrain of contestation of the meaning of organic out of the alternative agriculture social movement" (31). This means that rather than debate organic practices within a small, specialized social movement, organic methods became more broadly accepted because of certification. Certification allowed farmers to tap into global organic markets, and the exporting companies desired high standards. More recently, there is talk of developing a domestic organic certification scheme that would be more closely aligned with the organic grassroots movement, as many growers still use the "on-trust" or noncertification system domestically. Overall, this article shows the complex circular process "by which contexts, constructions and processes of circulation" (34) influence the evolution of organic production in New Zealand.

Discourse analysis is a research technique that involves reading and finding themes within public and private documents, interview transcripts, and historical books, pamphlets, and magazines. This approach was also used to analyze the organic movement in the UK (Reed 2001). The focus was the Soil Association, an organization whose influence began with Lady Eve Balfour in the 1930s and continues to the present; it now certifies 70 percent of British organic farms. The highs and lows of this organization were tracked within the context of other political and environmental issues. At one time shunned by the government that sought to facilitate conventional methods and the Green Revolution, the Soil Association has now become the British government–sponsored supplier of organic farming information. How times change! Apparently the mad cow scare put the issue of food safety on the public's radar screen, and the Soil Association has thus been able to inform people about the benefits of the organic farming alternative. Perhaps similar consumer concerns will ultimately lead American consumers to demand higher governmental accountability for the U.S. food supply. Likewise, consumers seeking alternatives may stimulate the continued growth of organics.

Deeper theoretical issues of how organic farming meshes with ideas of modernity were described by Kaltoft (2001), who says that organic farming has been a successful recent social movement. She draws from the case of Denmark to investigate how "institutionalization" of organic (in the form of national standards) is a form of modernization. What is useful about this article is that the author always links the theoretical notions back to real farmers and shows how they fit into the broader philosophical discussions. Premodern societies were mostly subsistence based with no given separation between culture and nature. Along with industrialization came the modern era. For our purposes, the key thing is that science became the "privileged form of access to knowledge – especially knowledge about nature" (147). Postmodern (or reflexive modern) is characterized by questioning and realizing there is not simply one way to understand nature. Individual organic farmers fit into all three of these categories – more premodern lifestyles versus a scientific approach to organic methods versus thoughtful, intentional choosing of an organic farming lifestyle. Add to this theorizing the issue of how national organic certification standards (institutionalization) influence the modernization of organic farming. Kaltoft lucidly describes these complex ideas with a bit of humor: "Going beyond modernity means to a certain extent becoming premodern again, but in the knowledge that we now live in hundreds of Middle Ages at the same time" (156). This means that organic farming is still linked to the premodern ideas of unity in nature and culture, but at the same time is postmodern in the sense that there are numerous human and nonhuman entities that are mobilized to create knowledge and technology in the organic movement.

Another study delves into the notion of sustainability in organic farming (Rigby and Cáceres 2001). These authors struggle to define sustainability (it is obviously complex) and explain how organic agricultural methods (also complex) may fit within this broad concept. They say that "what is a sustainable technique will vary both temporally and spatially" (23). True, but unfortunately they never emphasize the key point: all forms of agriculture are unsustainable. Organic methods may seek to do the least damage, but any agriculture upsets the balance of nature. They note that research has pointed us in opposite directions: organic methods are the only form of sustainable farming versus claims that organic methods cannot be considered sustainable. Organic farming is the only form of sustainable agriculture that is codified with specific written regulations, but these standards only refer to production methods and not to issues of social justice, ecological responsibility, or rural sustainability.

It would be nearly impossible to assess these philosophical issues within a certification system, as these vary among individuals, cultures, and nations. There may be social goals within an organic system, but these cannot be judged through production standards. One interesting question arises: how can we justify the sustainability of organic farming as it exists within our overwhelmingly unsustainable society? Can we expect organic farms to isolate themselves from the rest of America? There are also intriguing theoretical questions: how open or closed can an agricultural system be (energy, markets, inputs) and at what scale (farm, region, national) are we assessing sustainability (Rigby and Cáceres 2001)?

Allen and Kovach (2000) ask whether it is possible for the organic market to contribute to progressive environmental and social goals. "Green consumerism" is where people become informed and "vote" with their shopping dollars for products that are more environmentally friendly. Organic products are specifically labeled to draw these sorts of green consumers, but are they more environmentally benign? Yes, say the authors, in the current organic production system. But this could change as organics become more concentrated and competitive like conventional agriculture. Allen and Kovach note that organic standards cannot include philosophical concepts of ecological balance, so the holistic basis of organic production has become segmented into component parts: inputs, methods, soils, crops, markets, etc.

Next, Allen and Kovach (2000) introduce the topic of "commodity fetishism" as a means for the market to change social relations. Commodity fetishism occurs in a capitalist system when the social relations that went into producing a commodity are concealed when it is sold. Most products are in this category. Rarely does anybody really think about the working conditions, salaries, property ownership, or labor fairness when they buy an item. According to the authors this "hides the source of profits and, therefore, it deadens social action and resistance" (226). To specifically address agriculture, they note that "defetishization" means to make these social and ecological relationships crystal clear, and organic agriculture can greatly benefit from this openness. Thus marketing of organic products clearly displays the unique aspects of organic farming, while conventional products seek to hide their use of pesticides and their high federal subsidies. You don't buy a box of conventional corn flakes that advertises: "This cereal is made from GMO corn that was produced with pesticides that polluted local streams; and we only paid the farmer a quarter, but we're charging you $3.99 for this box of cereal!" So organic marketing has effectively used defetishization (or making explicit the information about organic production) to woo consumers. But organic advertising only tells you part of the story. You don't hear about the large-scale organic farms or the big corporations producing organic goods. Still, defetishization in organic products could encourage real social change: "This transparency – reaching into the farm, the scientific laboratories, and government agencies – could contribute" to social movements and improved environmental management (228). In conclusion, the organic food market can improve ecological sustainability in agriculture for the short term, but Allen and Kovach caution that long-term change in agriculture will require changes in social, political, and economic systems. Can organic farming do that? Not just with a marketing label. There needs to be broader social action.

We should also take issue with the concept of organic food as a niche market. Organics should be available to everybody, instead of just the wealthy "niche" consumers. Food, and organic food specifically, must be removed from the market commodity paradigm that treats it like any other item that earns capital and expands profits. Agriculture needs to be "reunited with its social context" through locally connected food systems that include CSAs, food co-ops, and farmers' markets. We can use our knowledge of organic farming "as a tool to decentralize and democratize our food system." If we relegate organic farming to mere niche markets, we are not really creating an alternative system at all (DeLind 1994, 147).

Has the booming market growth in the past decade propelled organic food beyond niche sales, or is it still a niche market, albeit a larger niche, that is available to wealthier people? The fact that more urban farmers' markets are present now and many accept food stamp coupons may be a positive sign that organics may reach some lower-income consumers. However, most specialty stores and supermarkets seem to have price differentials that keep organic food out of the reach of many. On the other hand, as college students tell me, it is also a matter of priority. They claim that although they have little money (living on student loans and minimum wage jobs), they choose to buy unprocessed organic foods (like rice and beans) that make several meals for mere pennies. This relates to education. Someone must know what organic farming is all about and intentionally decide to purchase this type of food.

Retailers tend to view the stereotypical organic consumer as health-conscious and wealthy, "despite the much wider consumption base identified for organics through survey techniques" (Lockie 2002, 289). Unfortunately, by pigeonholing organic shoppers, the food industry is setting up artificial barriers that may have a long-term impact on organic food consumption. Instead, a "relational approach" in which research considers production and consumption holistically would be more accurate in characterizing organic consumers.

It is important for us to discuss these social issues and to question what we expect from alternative agriculture and organic farming, but most farmers don't have time for such discussions. They are first and foremost a business –

profit is the difference between being a farmer and losing the farm. Organic farmers are overwhelmingly busy seeking information, making management decisions, producing crops and livestock, seeking marketing channels, educating consumers, and planning for next year's crops. Many authors have simply assumed that organic farming is inherently a social movement. Organic farmers, for their part, would be less philosophical and most likely to respond: "I do what is best for my farm."

4

Organic Farmers on the Ground

A healthy farm culture can be based only upon familiarity and can grow only among a people soundly established upon the land; it nourishes and safeguards a human intelligence of the earth that no amount of technology can satisfactorily replace.

– Wendell Berry, *The Unsettling of America: Culture and Agriculture* (1977)

rganic agriculture has, to some extent, attained a new level of recognition that propels it beyond much of the previous literature. Trying to link past research findings to current organic farmers is a challenge because many studies are tangential to the complex on-the-ground actions that determine whether a farmer will be in business next season or next year. Farmers make management decisions based on economic factors, physical environmental conditions, and personal aims and attitudes (Ilbery 1978). Moving beyond the existing literature, the farmers themselves tell fascinating stories about place, farms, and innovation.

LEARNING FROM REAL ORGANIC FARMERS

I interviewed five certified organic farmers because they provided diverse examples of farm types and geographical variations within U.S. organic agriculture. I intentionally avoided both hippie granola and huge agribusiness examples, since I wanted to feature the profitable medium-sized family organic farms that form much of organic production today and much of our hope for organic production in the future.

This chapter introduces the farmers, their farms, and their perspectives on organic farming. Given what these farmers teach us, the next chapter provides an overview of the key factors that influence successful organic farms.

Location of Five Interviewed Farmers

These farmers are some of the most colorful people I have ever met. Each in their own way has stood up against incredible odds and worked to develop a successful organic farming business. I visited and telephoned each farmer multiple times between 1994 and 2004. I taped all conversations and typed out the transcripts. My initial interviews were open-ended, as I wanted to learn as much as possible. Later I often called or e-mailed the farmers to chat or to ask specific questions: Has the early spring rain been bad for you this year? Did you decide to rent that extra land? Did the vetch cover crop work out? After I had all the written transcripts completed, I reviewed them extensively, and I listened to the taped interviews repeatedly. This helped me discover trends, identify themes, and pull out key topics. Each time I read an interview, I made lists of key ideas and scrawled over the lists with new ideas and topics the farmers taught me. Rather than using numbers as data, I used the farmers' words as data. Essentially, I used qualitative data analysis techniques to discover the key factors that influence these geographically and operationally diverse organic farmers. Theoretically, I am motivated by a pragmatic behavioral approach that seeks to understand the context of agricultural concerns by investigating farmers' experiences (Duram 1998b).

ORGANIC FARMERS IN FIVE GEOGRAPHIC REGIONS

Let me introduce these organic farmers. In upstate New York, Steve Porter and his family have shifted their operation from conventional livestock to wholesale organic vegetables and a successful Community Supported

Agriculture (CSA) program. In Florida, Mary and Rob Mitchell produce high-quality organic citrus; they have also developed a successful packing house and tapped into national markets. In Illinois, Joel Rissman works land that has been in his family for generations; now he and his wife sell their diverse mix of organic crops and livestock through multiple marketing channels that include direct sales. In eastern Colorado, the Bensons are fourth-generation grain farmers who are innovative in their organic crops, equipment, and marketing. And in California, Phil Foster operates a busy organic vegetable production ranch that has shifted its focus to local and regional sales (see map 1).

CSA WITHIN AN EVER-CHANGING FARM

Steve Porter, New York

Steve tells me that his farm is "halfway between Buffalo and Rochester. That is one important part about running our CSA – having a lot of people near by. We are close to two fairly large cities. Each of the metro areas is close to one million people." The CSA group is just one part, or the latest step, in this diversified five hundred acre farm that is run by Steve, his brother, and his dad. "I am a second-generation farmer. My parents bought our property here in the mid-fifties. It was basically a cattle-feeding operation for twenty years. Then we diversified and grew the crops we needed to feed the cattle, you know, hay and corn for grain and a little wheat for a cash crop, but not much. So it was basically a livestock operation. We added hogs in the seventies, bought feeder pigs, and finished them out. Then we built some farrowing and nursery facilities in the late seventies. And that is about where we stayed until the late eighties when we started adding sheep to the livestock enterprise. By then the cattle were very few or nonexistent some years."

Because of economic downturns in the livestock market, Steve's family began to look for other ways to diversify. "My dad had been reading about organics and the potential for it. In 1990 we started transitioning some ground. We had a little field we had always used for a garden spot, and it didn't have any pesticides or fertilizers on it, so it would certify. The first year we had half an acre, a big market garden. That is how we started, with a half acre." Prior to that, the focus had been livestock. "We never grew vegetables for marketing, so we were transitioning in a lot of different ways." Because of their crop patterns, their conversion to certified organic acres was fairly straightforward. "We started transitioning more ground and that was fairly easy because we had hay fields in our rotation, and those didn't

Steve Porter on his farm in upstate New York. (Credit: author)

get pesticides or fertilizers. So it was very easy to start transitioning ground in. And then we transitioned more and more ground into vegetables. From 1990 to 1996 everything was sold mostly to large wholesalers and a few small local stores in the Buffalo and Rochester area." The main organic vegetables they produce for wholesale are cucumbers, cabbage, winter squash, and onions.

The next transition was from wholesaling vegetables to more local sales. "We started the CSA in the 1996 growing season. And what happened is for years we'd been hearing about CSAs and didn't figure they were for us. Anyway, in the winter of 1995–96 we heard of another CSA more towards the Buffalo area that just wasn't large enough to support the family that was trying to run it. They got discouraged and went back to their real jobs and we heard about it. It was something we had been thinking about, but we were worried about trying to get one started from zero. We talked to them and they gladly gave us their mailing list database. We put a letter out introducing ourselves and saying that we would be willing to take over. Met with some of the people, the core group type people, and that is how we started." Thus the Porters were able to accept risk in the transition to

Steve Porter giving a tour of his farm. (Credit: author)

the CSA, but at least they had the basic framework of potential members to contact.

Steve guided the initial group's development. "The first year we had maybe a hundred families in the Buffalo area." Now the complex operational organization, which varies between the groups, is purely his own. "They were set up in what we call distribution groups. So they were used to one person driving out to the farm and picking up ten bags for their group and going back. That is how we run the Buffalo area people. We have about 120 families. I would have to go back to my database and start adding. It is about 110 to 120 families split up into nine groups. So they average about fourteen or fifteen families per group."

"Each group is usually a group of people who live in the same suburb or part of the city. They are split up by geography. One person from that group serves as a coordinator, who I give a discount to for some of their troubles. They set up a driving schedule for our season, and each week one person from the group drives out to the farm. The bags are already made up, so they just pick up the bag and maybe stay and visit. Then they go back and the same location each week serves as a pickup point. That is the same every week, so that is not a confusing thing. The coordinator doesn't drive every week. They all take turns. So over the course of the season I should see every person in the program at least once, some twice or three times, depending on the size of the group." Other than this pickup, the CSA members do not work on the farm.

Steven explains that the delivery system is somewhat different for the other city. "Now in Rochester, since we weren't starting with anything, the first year we had maybe twelve families. They got together and one person drove out, and that didn't work out too well. The following year we started a delivery system, and we basically drive around to the different suburbs and the city of Rochester. We leave six bags here and eight bags here. Same type of program, but we do the delivering and we charge a little extra for it. One of my employees does it Saturday mornings. He just loads up his pickup and he leaves here at eight in the morning. He actually lives in a suburb of Rochester, and his home serves as one of our pickup points. So he gets home around noon. It takes him about four or five hours. We don't have quite as many families in Rochester – maybe about sixty."

All together, then, "We are right around two hundred because we do have some local people that pick up right at the farm. We call it a share." He has two helpers, "one in Buffalo and one in the Rochester area, who help me keep the database. I just send the information to them, and they keep the database and print out the labels to put on the bags. You know, we really don't need labels because the bags are mostly the same, but we put everyone's name on their bag and their group letter."

The CSA season "usually starts in late June for us." But, of course, it depends on the weather. "This year we had such a cold wet spring we were late starting. We didn't start until the first weekend in July and finished the first weekend in December, which is later than I would have liked but that is what we had to work with." Last year, they grew more than twenty-five types of vegetables. "I would have to sit down and add them up, but it is plenty." For the CSA weekly shares "we use a standard size paper grocery bag, and because at times we have things going into the bag that are moist or wet, we use a plastic bag around the outside of the paper bag so that if it gets wet the bag won't rot and have everything fall out." In terms of vegetable quantities, "well, that depends on if it is heavy dense things like potatoes or squash where it is heavy but not quite full, or light and fluffy things that don't weigh a lot but make the bag look full. So we try to put as much different selection in the bags each week as possible."

The cost of the Porter's CSA shares is reasonable for organic produce. "This year it worked out in the Buffalo area people who picked up their own bags was $11 a week. I think we were at $230 or $240 if you signed up early and $250 or $260 if you signed up later in the season. And the Rochester people, since we deliver the bag, that is an extra $30 a season. We get a little over a buck a week to deliver them. I am not making any money off it, but at least I am getting paid something for it." Plus Steve gives incentives for signing up new customers. "We also offer a referral fee. Say you are happy with the program and you have a neighbor that saw you getting it and they are interested. If you sign them up, I'll knock off $10 of your sign-up fee."

Steve is honest about the economic benefits of the CSA. "Like I say, it does add some management challenges for us, but there is good margin in it. You know, I won't try to schmooze ya. It will make us money." So he is trying to increase their membership, based on understanding the members' needs. "In a few weeks I will be sending out what I call a year-end letter and a survey that will give people my take on the season last year and maybe any improvements we are thinking of doing in the coming year. It also has pricing information for the coming year and we put in a survey also. We even put in an addressed return envelope, so people can send the survey back without having to look for my address. We probably get 50 percent of the people to send back the surveys; people are pretty good about that."

To summarize his feelings about how the CSA fits within the farm: "We like this program." Indeed, due to their organic certification and work in identifying markets, their farm size has increased. In terms of farm size, Steve notes, "Since we have certified, we have picked up more ground for growing grains, and some of the soil is good for growing vegetables. We have been up to five hundred acres in the last three or four years." Of that total they own about three hundred tillable acres.

Luckily, Steve is organized and has developed an efficient system for bagging and labeling the two hundred CSA bags. He downplays his system. "Well, we've been doing it seven years now, and we haven't changed the system much. We have a lot of people working here on our wholesale vegetable sales. What we do on Friday afternoon my brother and I sit down and say, what are we giving the people this week. (We call them 'the people.') We make a list and give them two of this, one of this. And we just start sending people out to pick, and since we know how many people we are dealing with, we need eight hundred leeks because I want to give everyone four leeks. I'll tell the guys, 'Get a little extra, get nine hundred.' There is nothing worse than when we are bagging everything up, to run out just before we are done. After everyone gets their four leeks, there are maybe a hundred leeks left. That is kind of the perk for whoever drives out that week. That is my theory. If someone takes an extra six leeks, it is not going to hurt me financially, and it is good PR."

He continues to describe the system. "So we just have it down. We do that Friday: wash it, get everything prepared, and put it in a cooler. Then on Saturday morning, we basically set up a big buffet line with heavy things on one end and light, fluffy things at the other. Then everyone knows as you go down the line you get one of this, two of this. We have people serving the bags, so you walk down the line having stuff dropped in. At the end, either my daughters or I put the bags in the right groups. And we do a weekly newsletter, so we put a newsletter in the bag. We put a recipe in there and farm news, and people really like that." He is matter-of-fact. "So that is our Saturday. We can bag up all two hundred bags; with seven or eight people we can do it in an hour and a half. Once we do it a week or two, even if we have new workers in for the season, they get in the groove like everyone else. It is actually pretty easy."

Although this CSA does not require members to work on the farm, Steve is aware of the need for "PR," as he puts it, or the need to communicate with his members and maintain an open door. In fact, he says that the newsletter is a family affair at his house. "My daughter and I sit down late in the week and she does the typing because I type with one finger. They have grown up with computers. My daughters are fifteen and sixteen." They like to help out with the newsletter, and his younger daughter helps him run irrigation. As with most farm families in the United States, Steve's wife works off farm. She is a nurse. Her job is absolutely necessary "for family living and health insurance, those types of things. We don't have to use her salary to meet farm bills. This last year was a pretty tough one, weatherwise. It takes the pressure off. I don't have to worry about feeding the kids." As for other farm work: "It is my brother and myself; my father is still somewhat involved. He will be seventy-three this summer. He isn't doing as much as a few years ago, but that is expected."

While Steve enjoys the CSA and its direct marketing profits, the timing of the CSA is often at odds with other activities on the farm "because this is not everything we do. . . . Between the wholesale vegetables and the barley and corn and soybeans we have to plant in the spring, we don't try to hurry up the CSA season that much. Our main wholesale business doesn't start up until almost Labor Day." The farm is divided up among the vegetables and fieldcrops. "About 100 acres in mixed vegetables. Corn usually 100 to 120 and about the same for soybeans. Barley and other small grains would be 100 and the rest will be either pasture or hay fields or fallow." Their wholesale vegetable business is still in place, but the CSA has encouraged them to diversify more than otherwise. "I can tell you that quite a few of the crops we grow I wouldn't grow if we didn't have the CSA. I wouldn't be able to sell enough of them to justify the time. But with the CSA and the better profit margin, you can justify doing them. Growing these things helps our local sales to the small Co-op type stores since we grow things like eggplant or peppers or different tomatoes. Oh, a couple of years ago we tried leeks, which are a relative of the onions, and they proved real popular."

As he outlines his future plans to expand, he notes that the main limitation is time. "As I say, we are really trying to grow the CSA. Well, a lot of people call and ask questions in the spring, and I will come home at night and the answering machine is full." The busiest time is "in June when we are still planting things and still cultivating, and I'll think, okay, we have to start the CSA in two weeks. I have to get all of the database information to the people who do it for me so they can get it organized and printed out for me. There are a lot of things I need to do, and sometimes I hope it rains so I can get it done and not feel like I should be in the fields working." He continues to describe crop planting and timing. "There are only so many hours in a day. With the CSA you want to have fresh green beans throughout the season. So you want to plant them and when they come up, plant some more. Well, you come in the end of the day and you look out where you are planting green beans and say, oh they are up, I have to get some new ones planted. The tractor I need is in the wrong spot, so I have to go get that equipment. The actual planting takes five minutes. You just go down the field with a four-row planter. But the prep time could be a couple of hours. By the time you get everything organized to go, it is an hour to two hours." And they grow over twenty-five types of vegetables, so all these activities add up.

In terms of hired labor, they have seasonal employees (fifteen or so, at the peak of harvesting in August) and one man who does the CSA delivery and manages the greenhouses. They have built three greenhouses. "We have four to five thousand square feet of greenhouse space. Well, we grow all our own plants. It is not that we really want to, but for organic certification we had to find certified plants, which wasn't easy. So we've been forced to build some greenhouses. My brother manages all the help and the livestock. I do all the crop production and all of the mechanical work. So we have a pretty good division of labor. We are in a pretty large vegetable growing area here in western New York, so a lot of Mexicans are heading up here because they know there is work. So finding people hasn't been a major problem."

New this season, the farm will employ a married couple to help with field and office jobs. Steve thinks that having a woman in the office will be especially helpful. Women "seem to be more comfortable dealing with another woman. And I think having Kathy here to meet with people and talk over the phone, they will be more comfortable. It is going to take some time to train her, but I think it is going to pay off for us." Plus having some additional field help and management will save time. "That is what we are hoping. At the morning meetings I can say, 'Ok, we should get this, this, and this planted and get this cultivated, run some water over this.' That sort of thing. So that will be his day. And I would have tried to cram it in a few hours, and some of it just wouldn't get done because I just don't have the time." Steve is also proficient at building specific planting and harvesting equipment to fill specific needs of their diverse cropping systems and the smaller fields of organic vegetables.

Steve is definitely trying to expand the CSA, but needs more time to advertise. "We started with a hundred families, and it has been growing slowly and steadily. Right now it is kind of stuck around two hundred families. I haven't been actively promoting it. I have mostly been doing word-of-mouth and these guest bags. It's one thing we started doing a few years after we started. During the week you can give us a call and say your neighbor wants to try it for a week. We will gladly put up what we call a guest bag, and that person can try it for a week on us. They pick up the bag and see what it is like driving out to the farm. It is not hard, but it's not going to the grocery store like you are used to. We give them a sample of what we produce, and they can try it and see what they think. If they like it, give me a call and I will give you a price, pro-rated, for halfway through the season. We get you organized into a group and away we go. I probably add, over the course of the season, 5–10 percent to our program. It is cheap advertising, and it does work."

The biggest barrier to more effective advertising is, again, time. "That is what usually gets us into trouble in the spring. I have always had to try to organize the CSA while doing everything else in the spring. That is one reason we don't advertise much. I feel the best time to advertise this program, to get people interested, is the few weeks just before you start. Then put some general ads in and get people calling about the program. Right now is when I have time to talk to people, and who is going to look at something that starts in months? They will look at it and say, 'That looks interesting. I'll call later.' Then it gets lost in the shuffle."

While he would love to shift completely to the CSA and give up vegetable wholesaling, Steve notes, "Realistically I don't think that is possible. We would love to do four hundred, five hundred families or more. It would be great if we could get to that point. But I would probably still do the wholesale because we have identified some niche markets that we feel we can fill," particularly to natural foods stores in the Boston area. In terms of diversifying the entire farming operation, "there are some years that the weather is bad or the vegetables don't market, or there's competition from other areas. The livestock is a steady income." The livestock are not certified organic because "we haven't identified a good enough market." He has tried to dovetail the meat sales with his CSA, but with little success. "Cattle, I am down to two steers, which I have been trying to sell through the CSA. It has been a little underwhelming. I was somewhat disappointed." He says he knows several organic livestock producers across the country who are "just not getting enough of it sold and are quite discouraged." One friend in Iowa, for example, "is putting together good quality organic heifers and steers and can't get them sold." On the other hand, Porter's livestock focus is now different. "Our main livestock are the lambs. We are selling them direct, locally. There is enough of a Muslim population around that loves their lamb." They have five hundred ewes and are hoping to expand.

The livestock portion of the Porters' operation is responsive to market changes. "We are quite flexible. The barns we use for our livestock are flexible as to what they can be used for. The sheep use just an open shed. It isn't concrete and permanent. We used chainlink pen setups. We can land the ewes inside or outside or put feeder lambs inside. We are quite flexible in what we do. If the buildings sit empty, it's okay. They are not real expensive." Steve describes the highly concentrated hog industry, in which smaller producers (with under five thousand head) have fallen out rapidly. "We got out of hogs in the early nineties. We saw the writing on the wall, which way the hog business was going. Don't regret not having them around at all. There is no real market to sell hogs to anymore. You would have to sell them to the big packer, go down south. We just didn't see the hog business going any way we wanted to go. We just didn't want to be tied into it year after year. You have to have the right facilities so you can produce the hogs efficiently enough so you can hopefully eke out a profit. Those eight-cent hogs a few years ago . . . we have no regrets being out of the hog business." On the other hand, with their sheep: "We have market opportunities that we somewhat control; not control, but are happy with. And if we are not real happy with our local sales, there are a couple of companies in Pennsylvania who are somewhat competitive."

Steve describes agriculture in this region. "We are in a fairly good dairy area and cash crop area. Mostly processing type vegetables: green beans, sweet peas, and sweet corn are the three big ones." He says that "the biggest farms in the state are centered within about a five-mile radius around us. One of the biggest vegetable farms in the country is right next-door to me. They farm, between three or four counties, ten thousand acres. There is one farm near us that is milking three or four thousand cows on two locations. So for New York State agriculture, that is quite big." Even though farm size is increasing, Steve knows farm profits are decreasing. "Operations are getting bigger and bigger, and running more ground and more ground because there is hardly any margin in what they do." He thinks the future for conventional growers is bleak. "I have a feeling that processing vegetables are on the way out. The prices keep going down that they are paying the farmer. What they could make money on years ago, for yields, just doesn't cut it anymore. The prices are so low." He gives a specific example: "There are a lot of onions growing in our area, and I know a lot of farmers selling yellow cooking onions for a dime a pound, plus or minus, and you see them in the store for 79¢ a pound. Somebody is making tremendous money on them."

And yet the conventional farmers are not willing to change. "They look at the organics and what we do, and they figure it is too much work, so why bother? Oh, they grow the genetically engineered soybeans, and it is so easy. Why go out and cultivate? If that is the way they want to think, I don't encourage them to think otherwise." The Porters' diversity, organic certification, and organic direct marketing are what keep them in business. Steve says that the CSA is much more profitable than wholesaling the organic vegetables. "We are up to a buck a pound with the CSA, give or take, and a lot of the vegetables go out of here at twenty or thirty cents a pound for wholesale." Over the years, this farm has evolved and changed focus and even earned organic certification, while neighboring conventional farms have failed. Steve sums up the difference: "We haven't been afraid to change what we are doing." But why is that? "I don't know. We got tired of banging our heads up against the wall with the hogs and the cattle. We figured we had to do something different or not be here. Part of it might be financial stresses."

Their most recent answer has been the CSAs. To increase the presence of CSAs in U.S. agriculture, Steve thinks that the key issues are buying local, accepting seasonal foods, and trying a variety of vegetables. "I think if people knew where some of their stuff was coming from . . . ya know? They go in the store and see vegetables that look nice. If they knew it was coming from overseas, they might say, 'Oh, I'd like to buy local.'" In addition, "people have gotten so used to walking into a grocery store any time of the year and buying exactly what they want. They are used to walking in the store in the middle of winter and buying a green pepper. That is hard to beat." He knows from his member surveys each year that "a big reason for our CSA turnover each year is people are not used to cooking with different things. That is a big challenge."

This is where Steve's communication skills help out. "In our newsletter, we make sure there is a recipe so they know what to do with it. Sometimes we go as far as identifying things in the bag. Well, especially new members, they may have never seen that vegetable before in their life. Say, like kale or collards, a lot of people don't use them. And maybe some of our peppers, if we throw in some hot ones, we'll say, 'Be careful these are hot,' so they don't go and bite into one." The fact that their farm is certified organic "is quite important. I think it is a good selling point, but I have never asked that in a survey. I would say most of the people we are getting in the program are better educated, a lot of younger families and I think they are very concerned with what they eat, especially for their children. And a lot of people like bringing their kids out and showing them that food doesn't just show up at the grocery store, it has to be grown and all of that."

This is a busy farm producing vegetables for CSA members and wholesale, grains sold regionally, and livestock for direct marketing. They also grow some melons, and for now "that is the only fruit we grow. I think having some fruit like apples or pears in the program would go over huge. I have no fruit-growing experience, so it is something to learn." It sounds like this farm will continue to evolve, with Steve and his brother at the helm, following in their father's footsteps. They are managing well today, while looking toward the future.

ONLY THE BEST CITRUS

Rob and Mary Mitchell, Florida

As a visitor drives into the yard of this fourteen-acre citrus grove, two dogs run up in greeting. Rob is zipping around on his fork-lift, filling a semitruck that is waiting with its engine idling. In a short time, another semitruck drives in – it's a day early. Rob starts yelling about it and Mary yells back, telling him to switch out parts of completed pallets so they can fill this early truck with citrus now, then they can make up these orders once the workers come in the afternoon. The place is hectic. Both Rob and Mary are high energy, strong-willed people, so yelling is common. They are proud of their delicious citrus. Mary says, "In terms of organic citrus; we wrote the book!"

Rob Mitchell picking his Florida citrus. (Credit: Jon Bathgate)

Mary Mitchell shipping out an order of Florida citrus. (Credit: author)

Rob explains, "This is the oldest organic gig in the United States as far as citrus goes. I am pretty sure there is nothing close in California. This has been a grove since 1879 and it has been organic since 1946. Strictly organic."

Rob explains how it all began. "My brother and I came here in 1975. Mr. DeWolfe owned this place. He wanted someone to work *with* him and not *for* him. He decided to sell it to me and my brother because his family was going to sell it out as soon as he was dead and gone. So my brother bought the other side of the road, but he lost that in his divorce." Of course, it hasn't been easy. "It's been a lot of learning by doing and then we froze in 1983, and then we went about five years without having any grove at all. And then I started planting again in 1988. Then in 1992 or so I started coming into production and I haven't stopped since. It seems like yesterday. I wish I had fifty or a hundred acres right in this area, right on this Fruit Land Peninsula."

Mary is from Kentucky. She met Rob and followed him to the farm in 1987. She explains how the grove takes the best of both of them: "It's not a job. It's fun. You know, we enjoy packing it, the people. I love it, so you know it's not a job. It's fun and we get paid for it. It's got to be in your blood, that's for sure. His thing is growing and maintaining; my thing is marketing and collections. He doesn't like doing that part, and he is not as good at it as I am, and I don't like going out there and hoeing trees, and he does, so it works out perfect."

Rob points to a box imprinted with their brand name, Eagles' Nest Grove. "Right there it says tree-ripened fruit. If I don't eat it, I don't sell it. If it is sour, we don't sell it." Mary agrees. "We don't sell a single orange until we sit down and eat six to twelve of them apiece. And when they are really good and we run out to get more to eat, that's when I pick up the phone and tell people we're coming online." Rob explains, "We don't pick and store. You don't call here and say I need this, this, and this tomorrow. It don't work like that. You give us three or four days because we have to pick it, pack it, and then it is ready. Do you want it fresh, or do you want it three weeks old? How do you want it? We deal in fresh fruit. Now all our competition, most of them pick and store."

Rob and Mary describe how most citrus growers try to sell their crops early, to get a jumpstart in the market, but taste is sacrificed. Synthetically formulated ethylene gas is applied to fruit that is completely green in order to ripen it. Rob says, "Oh, they will have them on the market in September. They will be emerald green on the tree, they'll gas them, and they throw some color to them. They may look good on the shelf, but you go buy them and you won't be in any hurry to go buy more. Growers wonder why they can't make any money. Orlando tangelos, if you ship them in November, they are like battery acid; if you wait until December they are pretty damn good. We have a lot of markets, like in Chicago and the Northeast they can't give away Orlando tangelos. I said, 'Well, you bought them from the competition. Organic Orlandos, if you buy them in November, they ruin the market. Everybody knows if you want a good Orlando tangelo, you come here. You get the best.' " Mary agrees. "We have the best product in Florida. Because we hold it until it is naturally sweet and tree ripened and the sugar content is there. As far as selling to grocery stores and wholesalers, we have the USDA #1 product. Not only does the interior have appeal and color, the exterior has the color. You know a lot of growers down here say it can't be done. Well, yes, it can."

Rob proudly says that the only citrus he knows is organic citrus. "People say, 'You have a niche market.' I say, 'I didn't know there was anything else. That is all I know. I didn't know there was any other kind.' People start talking to me about chemicals, and I am like, 'What is that?' I heard of these things, but I don't know anything about them. I don't want to know." They grow eleven varieties of citrus (various oranges, tangerines, and grapefruit) on fourteen certified organic acres. Much of the citrus is sold to distributors for major natural foods grocery stores throughout the United States. Plus a small part of their business is mail order or "cash in yard" as people stop to buy a box of oranges.

The direct marketing component of their business has been shrinking, to their relief. Mary explains, "Most of mine go out of here right on these trucks. That is the market I want. Because with mail order it is feast or famine. The customers may be able to afford it one year and not the next. Plus the extra handling. Really when you figure it out, we are making more money wholesaling. Because you don't have the extra time, you don't have the more expensive carton, and you don't have the shippers banging you for $5 if you misspell a last name or if your customer gives it to you wrong." Rob, as always, is more blunt, but still notes the pros and cons of direct marketing to consumers. "We used to do a ton of it. I always liked it because it is just another avenue to get rid of fruit. But it has its advantages and disadvantages. One of the advantages is 'you send us the bread, we send you the fruit,' but you have a million phone calls to deal with. Somebody is always going to complain, 'I got a dry orange. I had one bad piece of fruit.' It can be a real pain in the ass. Everything has its good points and bad points. I have always liked doing the mail orders because they are the only ones who call you and say, 'That is the best fruit I have ever had. Send me more.' You get more compliments, and then you get some pain in the ass things to deal with, too." In any case, Rob says that "direct mail, I'm going to say, is about 5 percent. That and cash in the yard. City slickers who have a lake house down here. They eat 'em up."

As with all successful organic farms, marketing is key for Rob and Mary. And they have come a long way. Mary explains, "Oh, gosh, when we started, nobody would help. They would not tell you anything about who the buyers are and how to go about it. And then one night we saw something on the news, and they mentioned Whole Foods, and we were like, where is that? So we got on the phone and started calling around until we found one in Maryland. So I get ahold of them and find out who their organic buyer is. He told me about other stores in Boston. Then we go through another wholesaler in North Carolina, but the distributorship is Georgia, North, South Carolina, Maryland, New York. They're all over."

When she's dealing with the big distributors, she has learned to stand her ground. Nobody pushes her around. "Now I get right back in their faces." Distributors try to change truck schedules and price quotes at the last minute, and she really has little flexibility. She describes one recent example of a conversation she had with a distributor. "I said, 'We are growers, we are little fish, but you wouldn't be that big fish, when the little fish are gone, and you need to remember that.' And I said, 'Is the product good?' He said it was going like wildfire. I said 'Okay, you give me your order by noon and we will load your trucks after 4:00 Mondays and Thursdays. Don't start this attitude problem. If a grower in California just ruins your day, don't you think about calling me and ruining mine because I will make it ten times worse.' He said, 'OKAY.'" After she traveled out of state to visit with a distributor, she walked out of a meeting with him. When she talked to him later on the telephone, she told him, "I am not playing with you, buddy. I do not have time for your little games. I know you have in the back of your mind that you are going to make me toe the line. It doesn't happen. Let's get this nonsense out of your mind. He said, 'Nobody just walks out of a meeting.' I said, 'I do. I've done it before, and I'll do it again.'"

On the other hand, most of Mary's marketing work is harmonious. "You build relationships. You build a business relationship, but it is a friend thing. Integrity. And once you build those relationships, you really don't have any problems marketing. They trust you. A lot of organics is trust, and it always has been. Some of them don't care as long as you say it is organic and they have that paperwork. But 90 percent of the people I know, they want the real thing and they do care. They care about their growers. I have them calling me saying, 'I wish I could do something for you. Are you worried about the freeze?' I have people calling me from all over the country, and that is very unusual. And they really were concerned. So, I mean, you really build up friendships, and they last."

Mary explains how they learned the importance of using their brand name. One distributor "bought a little bit and it flew off the shelves. He is the one that said, 'You have to have brand recognition.' I was using generic cartons, so he got us to print up cartons. In bold letters it would say Eagles' Nest, and it had a list of what we were coming on the market with. It would just fly off their shelves because people knew it was consistently quality." Relationships and word-of-mouth have truly built their marketing avenues. Mary says, "Brand recognition really put me over the top. People would call that distributor, and he would tell them, 'If you can get this girl's citrus, buy it. She is the best show in town. You need to buy this.' So he really helped me out a lot." And customers return because "they were all consistently happy, and they all tell me price is no concern. As long as the quality is there, they'll pay what you are asking. Because price per acre I probably make less than these other groves. But that's okay because I am happy with what I sell. I am very proud of my product." Rob says, "As far as all the organic fruit in the state, we ain't the cheapest, but we've got the best. And we also have the best pack in the state. When we sell a box of #1 quality fruit, I guarantee, it not only eats good, but it looks like a million bucks." A tour substantiates this claim.

Rob drives around the grove on his ancient tractor that starts most of the time, pulling a trailer that is loaded with wooden crates of oranges. He stops, picks a few oranges, and slices them with a huge knife. He gobbles half in one bite, neatly spitting out the seeds. He has clearly had plenty of practice at eating citrus. He says, "Damn, these things are good, and the trees are only two years old, and to put quality out like that. I wish I had ten more acres of them. Look how nice they are. They are basically pretty. Pretty. All right, we have one more stop." We continue to the other edge of the grove and then back to the packinghouse. Rob asks, "Do you want to see big? Look at these grapefruit. And good. A seven- and a nine-year-old kid was here yesterday with their granddad. I always give the guy a lot of fruit for his cows when we can't get rid of it. You should have seen those kids wiping out that grapefruit! And kids don't usually eat grapefruit. They were like, 'Wow, these are good!' "

Rob describes their concern for quality and their motivation for a high "pack out." That is of all the citrus that is picked, how much of it is fit for boxing and selling, versus how much is wasted (or donated, as Rob says). "To find anybody that is really truly committed growers, like us, they are far and few between. People that take care of the trees. We are in the fresh fruit market. Packout is the bottom line. The better you pack out, the more you are making, because basically everything you don't pack gets thrown away. Or donated." Mary says, "I'm the eliminator. I'm often the one doing the eliminations on the line. So our pack out would run 60–75 percent on most things, 90 percent on my tangelos. They are just a clean piece of fruit." Rob agrees. "Well, every variety is different. My Orlando tangelos, I run about 95 percent, which is super. My friend's grapefruit we are packing right now, he is probably packing 65–70 percent, which is super on grapefruits." Rob says that he was amazed at the poor quality of some of the organic citrus in the region. "There is a ten-acre certified block of organic grapefruits that belongs to a family that is inside of their 25,000-acre grove. The rest is juice fruit. This guy got in a pinch a couple of weeks ago, and we packed some of them for him against my better judgment. The grapefruit itself was not bad eating, but it was some ugly shit. He brought five hundred boxes up here on a flat bed trailer. That is fifty of those bins like that. I sent him back with about twenty. Terrible pack out. More than half. I figured we had about 37 percent pack out on that."

Mary says, "Florida has a bad reputation as far as citrus for not having eye appeal. I'm probably the one of a few who could pass the USDA #1 hands down every time. Because we have the eye appeal, and we get money there." Rob notes that if customers "are truly organic-minded, they don't care if it is not washed." But Mary says, "For a lot of the organic market it is not important, but if you want to get into an upper-end grocery store and get those repeated orders, it is important that you have that eye appeal."

Their fields are certified organic, and they have a certified organic packing facility. Mary describes how difficult it was to create this organic packinghouse. "The Department of Citrus came out. First, they wouldn't let us have a license for a packinghouse because we couldn't fumigate, and finally we appealed to the Board of Citrus Commissioners, and they rewrote the citrus laws for organics for us. We are the ones who got that instituted – that we did not have to fumigate because we are organic. We did not have to wax because we were organic. So a lot of the rules were rewritten for organic growers." Explaining the packing equipment, Rob says they "bought all that new equipment about three years ago. We had a good year and had a little extra dough, so we bought new instead of searching around buying some old used junk. I could have went around and bought all kinds of junk packing equipment, but by the time you rework it, you are going to have the same in it as buying custom." It is custom built for their operation. Now Rob and Mary grow, pack, market, and ship their own citrus, plus they pack, market, and ship for a few other certified organic growers.

Mary explains, "I have other growers right around here that I sell and pack for because it is good, it is quality." Although the citrus meets her standards for quality, Mary notes, "We pack for a hysterical group: a professor, a millionaire, a teacher, fireman, and a retired engineer. And then you throw Rob into the mix! These men will be the death of me yet! They actually tell me, 'Oh, I couldn't get picked 'cuz we went to a party instead.' Can you believe that? We don't even have Christmas around here, 'cuz we're so busy, but they just go off and don't pick! We pack for six or seven growers, it depends on who is on 'time out' – they are just like kids! – who's been misbehaving?! Then I don't want to deal with them for a while. I keep telling them, 'I've got orders and these people need their citrus!'"

Rob explains his philosophy: "I will pay premium dollars for premium fruit, but I ain't paying something for junk. I am at a point now where if I do any dealings with anyone I am not going to tell them anything. I am going to pack it first. I will pay you for the premium packed fruit boxes. The rest of it you come and get it out of here, unless I can sell it at a flea market or move it in the yard. I will work something out. Some of the grapefruits that came up here, what a joke. My packers are looking at me saying, 'What do you want us to do with this?' They have seen some ugly fruit, but not this bad. They say, 'What do you want us to do with this?' Just do the best we can and that is all we can do. We started packing them as Eagles' Nest, then I stopped and started putting them in generic boxes because I didn't want my name on it."

Rob notes that his certified organic packinghouse is different in terms of action and regulation. "If you look at a regular carton of fruit, it will say 'fungicide' on it. And that is why I am certified organic here. If somebody packs conventional fruit, they have to apply fungicides to the fruit as it is going through the line. That is the deal. A fungicide to keep it from rotting." But Rob feels that their fruit stands up fine. And Mary knows from her customers. "My fruit has a longer shelf life because it is highly mineralized. Due to the components of the fertilizers we use, it has the minerals it needs. And of course it has been growing in clean soils since the 1940s. I have a very long shelf life. Solids, pound per solids, you're not going to beat us."

Mary explains that she is careful to maintain detailed paperwork, so that each box, bin, pallet, and truck load is clearly attributed to each grower. "I have my boxes designed so where it says certified organic, I stamp the number. I buy from other people. I buy their crops, pack them, and sell them. So I stamp in their number. That is for product identification. And then on my manifests I also write whose product was inspected. If there were three of us there, I mark what number was whose. Each product can be identified. If they have problems, we know who to fall back on. Nobody else is doing that, and this is how we are giving everyone free rein."

Mary and Rob are concerned about fraud and lack of integrity in organic citrus. Mary says, "People who have just recently been certified are in it for the money. They see money is being made because the conventional market has bottomed out. It's the money. So what happens is they get in the business, and twenty acres is a nice living, and they figure out real quick that you can do whatever you want and the chances of getting caught are very slim. Oh, we had one finally get caught. He was packing 70,000 to 80,000 boxes of fruit a year, but he only had the same amount of citrus owned as I do. And then he was going out and getting it from abandoned groves, but it wasn't certified. All the buyers would say things, and I would say, 'Well, gee, he only has twenty acres.' When he got busted, they all said, 'Why didn't you tell us?' Hey, I don't bad-mouth. That makes me look bad. I told them, 'Why do you think I kept saying that he only has twenty acres?' I said, 'Can you figure out what I was trying to tell you? There is no way this guy can be doing this. You should be leery of him.' "

Rob says, "The bigger the operation, the lesser the quality on the whole. I also know there are people in this state who are unscrupulous and will cheat all day long and we have already had one guy get busted, but the one that needs to get busted . . . well, for some reason he is in good with the law." There is a real problem, according to the Mitchells, with "organic by default," which is when groves are neglected, not sprayed, and thus somewhat organic, except that there are no positive actions taking place. Rob says, "These guys had groves that weren't making money, they weren't taking care of them, and they looked like shit. So it was very easy for them to get it certified organic, and that is most of the organic state. If you are just in the juice business, production fruit, you can let them go. But if you look inside my trees, they are clean, no intergrowth. If it's a jungle in there, it harbors all kind of vermin. For all these other people, it will come around again. What goes around, comes around."

Mary describes the current trends in organic production: "A lot of people are in it strictly for the money. They have no integrity. But when you get found out, you won't be in the ballgame long. People are leery of organics, and I don't blame them, because I know a lot of things that go on. There are a lot of things you can get by with, and people aren't going to know it. But when you do get caught, you are done, decertified for life. And it is also a felony in Florida. Every time I ship anything out of state, I have to sign a piece of paper that states the statute and the law. It's a felony punishable by fine and or prison term."

Rob questions the certification rules and national standards. "I'm all for organics, but there is a lot of rules that have been written that sure makes it a lot easier for the Big Shots to get into it. They will ruin it." Mary adds, "And I'll tell you one thing. I do not like the certification groups. They don't stay on top of their growers, they don't check their trip tickets, and they don't check their records at all. So what I've been telling them to do is surprise inspections. They said, 'Well, we can't afford this. We are a nonprofit organization.' I said, 'I don't care. Charge it off to the grower. Have them sign a piece of paper. So much an hour. Surprise inspection. I want someone coming in knocking on my back door that I don't know is coming. They need to come through here and go through your records, billing, invoices." The certifying agencies need to do more. "None of them check back. None of them. There is a lack of enforcement there. You get checked once a year and you are done, you're free to do whatever you want to do." Finally, she hits the nail on the head. "Well, you see a couple of the big ones get busted and people hear about it. And that gives us all a bad name."

During the busy season, their lives revolve around picking, packing, and loading the citrus. Rob says, "For me personally, it's December. We were running sixteen to eighteen hours a day from Thanksgiving to the middle of January. Extra people: picking two to five and packing eight to ten. That's for my own stuff. What I am picking right now I saved just for the mail order, and I am going to be lucky to make it." Mary says: "we run ours starting in November, and other people we'll run until April."

Rob has to work off-farm for four months of the year. "My health insurance just went up to $508 a month, and my company pays for all that. That is $12,000 a year. That is why I work another job, just for health insurance. Jesus Christ, that stuff is out of control. I work for an engineering firm. Got me a full-time job. Only worked four months last year with a full-time package. That ain't a bad deal, is it? When I interviewed I said, 'Here is the deal: I don't go to work before May and on October 15 I am done, period, final, that is it. I can't work any other way.' They said, 'You are just what we are looking for.' Structural steel: water towers, power plants, nuclear plants, bridges. I like the bridges the best, though. They are wide open. It is usually six to seven days a week. When I leave here, I am not looking for a forty-hour-a-week job. I work four and a half months pretty much with only three or four days off. I came home for two weeks, but while I was on the job, you couldn't pay me to take a day off." This construction job takes him far from home. But he hopes to stay a bit closer to the farm. "Now we have about fourteen acres in production. If I pull a job in Florida for the next two years, I will have seventeen in production real quick."

As for pest problems, Rob explains, "This little bastard right here: rust mite. If you have a crop of fruit that looks like that one, you have a crop of losers. You can't sell that stuff out on the market." Conventional growers spray synthetic chemicals to kill the rust mite, but Rob asks, "Why would you want to spray motor oil on your trees? Vegetable oil works just great. I always fought rust mites, and I knew that there was something to do. I got this little gardening book, and it was talking about cottonseed oil, and I was thinking, why can't you just get a jug of Mazola and dump it on the tree? Then this guy from Peru said that they use olive oil, they have tons of olive oil down there, and they just spray it on the tree almost straight. So I got to thinking, why not use the vegetable oil? I use organic soybean oil. It is fairly thick. When you spray it on there it stays. It will go through quite a few thunderstorms. Now it won't hold three days in fourteen inches of rain. Nothing will. But just afternoon thunderstorms, it stays right on the trees. It has good staying power. It works wonders." Mary joins in the praises. "The trees were just a gorgeous green after we gave three applications. We basically tested it last year, and you wouldn't believe how gorgeous it is.

Our pack out rate was increased 15 percent across the board. Even our juice oranges were absolutely gorgeous. I had buyers calling me and saying, 'Hey, Mary, when did you go conventional? These are beautiful.' I said, 'Aren't they?' That soybean oil has done the trick." Rob continues to experiment, too. This year he may try a new pottery ash material to dust the trees and protect them.

They have distinct grove fertility management, as Rob explains. "All this weed over here is hairy indigo. Looks just like wood when it dries out. If you put that in your soil year after year, you will have nice, good-looking dirt." He contrasts his soil to conventional groves. "Half the groves you go to look like beach sand, no organic matter. Now all I have to do is maintain what I got. I can't add nothing to it. It's got everything that it needs." They also work with an established fertilizer company, Fertrell. "They are the oldest organic brand fertilizer in business in the United States, and they have the best in the world." Rob and Mary have had many visitors who are interested in their methods. "One day this one professor from the University of Vermont came, and he couldn't believe it. He was speechless. He cut oranges open; he took pictures from the inside, outside. He said he had been all over the world and had never seen a piece of fruit that had the exterior color, the interior color, and the eye appeal we have here on this farm. He actually was picking up dirt, sniffing it, and saying that was the best dirt he ever smelled. He said it smells like real dirt, and he was just amazed by the healthiness of my ground and of the piece of fruit. He was with the World Citrus Convention, and these guys go all over the world."

Rob describes some other visitors: "I had a bunch of people from South America out here a couple of years ago; university researchers and some growers. One guy went to the University of Florida in the fifties. He was a Dutchman from Holland but he moved to Peru years ago and had a nursery and grew pecans down there. The number one recommended cover crop was hairy indigo." He contrasts this to what the majority of growers use. "Hell, now everything comes in a five-gallon bucket, with a skull and crossbones on it. You would be surprised how many people come in here and say, 'Man, when are you going to cut all those weeds down?' But then there are some slicker that says, 'Whoa, buddy you've got yourself a nice stand of indigo.' They know what it is. Now the average tourist, they just think it is weedier than hell. They don't know nothing." Mary notes, "It seems like the bigger the organic movement gets, the more people come here. I think this year we had seventeen more scientists come here from Europe, Chile, Peru, Germany, and they had their doctorates. It was amazing. They were so excited. They had never seen anything like this in their lives. We have a big part of history here, but people who are just getting into organics, they don't even have a clue."

Even with their history and their colorful personalities, Rob and Mary have only succeeded due to sheer determination and hard work. Mary describes the difficulties they encountered: "When I got in this business, it was fighting tooth and nail. Nobody would divulge any buyers to me. We would have to read about them in magazines. We had to muscle our way in, and it was tough getting in. It took forever, but now I have a list of people who call me. People say, 'If you ever have excess, let me know. I'm interested in doing business with you.' I can't handle all the business because I have built the name for quality." Rob says, "I sell tree-ripened fruit, and I have no doubt I am going to sell my fruit."

DIVERSITY INSTEAD OF CORN

Joel Rissman, Illinois

This is the Corn Belt. The corn/soybean rotation rules here in north central Illinois. Joel Rissman returned to this area after receiving his college degree in agricultural engineering. "I graduated in 1991 and moved back to the farm. My dad was ready to retire from farming, and so I thought that, hey, I lived on a farm all my life and that it would be fun. You know, you can't get it out of your blood once it's in there. So I came back." He and his wife, Adela, started farming here on rented ground. "It's my uncle's farm, and it was only 372 acres, and I farmed it chemically for the first two years. I was pretty disgusted. You spend all that money, and all the chemical people and fertilizer people get paid first, and what's left over is yours. And there is just so little left over, and there is nowhere to expand around here. All the big farmers, if land becomes available, they'll put this huge rent on it. The little guys like me can't afford to compete with the big guys."

He knew that something had to change. "I had done some research, and it was in the back of my mind to go organic, but basically the first few years I didn't know how. The weeds were terrible, and I thought this isn't going to work. So I did some research and talked to other people, and I found out that it was possible." He began to seek out information from other organic farmers and publications. "I started going to talk to Paul. I was going to quit farming, but he showed me *Acres*. I called them up and got a subscription and ordered about $150 worth of books. And that winter, all I did was read.

Joel Rissman admiring the compost on his Illinois farm. (Credit: author)

And I read everything two and three times. And it was interesting because a lot of the stuff in that book was the same thing that I see happening on the farm. Some chemicals cause certain weeds to grow, and that is the same weed problem – weed chemical relationship – that I was seeing there. I said, 'This is crazy. I am paying for chemicals to get rid of weeds and they are causing *other* weeds to grow.' Everything in that book was pretty much like they came to this farm and wrote it.

"What really got me started was going to talk to my uncle, who owns the farm, and I said, 'I can't make it chemically. I am either quitting farming or switching to organic.' And he said, 'Well, don't feel like because of me that you have to use chemicals.' But I told him up front that it wasn't going to be easy, that he was going to have to get used to seeing a few weeds out there. He was raised by what I call the 'clean, green era' where you are judged by how many weeds are in your field and not by how much money is in your pocket at the end of the year." And, of course, organic fields may have some weeds, which was difficult at first, but his uncle "accepted it more and more," and the organic income was good. According to Joel, at least with some crops, "it was better than my chemical income. And we didn't have spectacular yields because of all the rain. So that is where I started. I told

The Rissmans at the Green City Farmers' Market in Chicago. (Credit: Glenda Kapsalis)

him it wasn't going to be easy, and it hasn't. But it is a lot more fun." Plus the chemical concerns are gone. "You don't have to worry about rinsing [after applications] and your kids getting into something."

Joel has clear opinions about the dangers of agrichemicals. He describes one neighbor: "Mr. McDonnell. He's retired now, but he farmed for years. And his story, he said every spring he would go partially blind. He came to find out it was from the chemicals he was using." But often it's the land owner, not the farmer who rents and works the land, who insists that chemicals be used because they think pesticides help to avert economic risk. For Mr. McDonnell, apparently, this was a concern. Land ownership and decision making are often distinct in conventional Corn Belt agriculture, as landowners often lease out land to farmers who grow corn and soybeans on thousands of acres of geographically dispersed ground.

When questioned further about his motivations for being an organic farmer, Joel says, "I thought about it in high school, because my dad has Parkinson's. It's from the chemicals. No one will change his mind or my mind. Yeah, he thinks that, too. It's from the chemical use and exposure to it. They get a neurological magazine that has found a pretty good correlation between chemical usage – insecticides mainly. And that is what he handled mainly. A lot of it was custom applied. But he always put on for corn rootworm, stuff like that." His father has since died.

Chemicals were just a normal part of growing up on a farm in the Corn Belt. "As a kid I can remember smoothing it off in the box with my bare hand and thinking, this stuff smells great. All we were doing is poisoning ourselves. And these 'lock and loads' are a farce. Sure you don't have to handle it, but you are still breathing it. You can still smell it. The vapors are escaping. What is worse – the dermal or inhaling the vapors? I don't know. But they are both just as lethal. I can smell it in our neighborhood when somebody is spraying. You can't see anybody, but it drifts miles and miles."

With these concerns in the back of his mind, Joel started transitioning to organic with 138 acres certified in 1994. And he reached 372 certified acres by 1997. "Yeah. That was about half of the farm first year, and then the whole farm." Joel describes his learning curve in organic methods: "In 1994 it was my first trial. We had a pretty good year. I did a lot of rotary hoeing like they said the third and seventh day. It was a dry spring, and that helped. I thought to myself, this is going to be easy. We had no weeds, and I had two hundred bushel yields just like everyone else, with less than twenty dollars fertility per acre. But then it wasn't so easy the next two years, with all the rain, and now lately we've had drought." The weather, as always, is a major influence on each year's success. "Well, some of our yields, I admit, have dropped, but I can't really say that it is due to what I am doing. Because one year we had twice the rain and the next year we had three times the rain as our normal amount. So I don't know if it was from weeds or just so much rain and late planting."

In 1999, his uncle sold off some acreage, so Joel now farms 300 acres, all certified organic. In terms of crop rotations, Joel explains his original actions, back in 1997: "You need small grains – legume-legume-corn-back into legumes and back to the small grain. So sometimes I get in maybe two or three years of small grain in the mix, but I try to go with a four- or five-year rotation. I raise oats, wheat, hairy vetch for seed, food grade beans, corn, alfalfa." Recently, however, this rotation has changed, as Joel now refuses to grow corn because of GMO contamination. "We try to raise the cleanest product as we can and there is no way you can raise noncontaminated corn, whether it is conventional or organic, because pollen can blow a maximum of eighteen miles. So how can you have a crop? And even the seed you buy; just because it is organic, doesn't mean it is 100 percent pure. There is no one in the industry that will guarantee a 100 percent non-GMO seed. So we have switched to grain sorghum instead." The advantages to the grain sorghum are that "it will not cross with corn hybrid. In fact, there is no one within a hundred miles of me raising it. And it works well in a drought, which was great for last year." He continues to grow oats, wheat, barley, and flax. "I started the flax in 1999. We raise the golden for human consumption, and my livestock is fed all of the brown."

The livestock have become an increasingly important component of their farming operation. "We had 1,200 chickens last year and we are probably going to do 3,500 this year. We sell to one restaurant that wants eighty a year." In addition, "We started the year with sixty-something head of cattle." Their organic cattle "are not getting all of the growth hormones, feed conversion additives, and all of those other chemicals. When you look at all of the chemicals, beef is probably one of the worse things that you can eat because of the contamination. It's a shame for me to say that as a beef producer, but it is the truth. I've got a book by this health guy. The first thing he says is get off all your beef, unless you can find a clean source. Then you can eat all you want. Because the grain they eat has chemicals in it, and then the growth hormones and chemicals. It is terrible. Do you know how they handle conventional cattle? It's terrible. They are constantly being fed drugs their whole life." He explains where he gets the livestock. "With the feeder cattle, well, the ones we currently have are from northeastern Iowa. They are organic. They came from an organic producer. They have to in order to be certified. Cows and calves have to be certified. And then to finish them out, everything I do has to qualify as organic." And there have been no problems with illness. "Not a one. I have never treated an organic animal ever." This contrasts with his past experience in conventional production. "Now conventional . . . oh, you don't even want to know how much money I spent treating them, and then they die on you."

And their chickens: "All of that is organic, too. We don't feed them any drugs. From day one, they get all organic feed." He describes their pasture-raised poultry: "We put twelve hundred chickens on two acres with the turkeys, too. We had about a hundred turkeys. They are inside for three weeks. You have to let them feather. Once they are feathered out, you can put them out to pasture. And then maybe another five to six weeks on pasture. Yes, we have a pasture seeded here. They are all on pasture. I think with the grain sorghum they will do better than with the corn." The logistics of processing and delivering the chickens has been challenging. "Well, the one restaurant wants fifty chickens a week," plus Joel needs packaged chickens for local sales and a farmers' market. "For processing, this year we will probably take these down by Kankakee or over to Trackside. He does processing of chicken, about a half hour away. I am hoping he can do it because the restaurant wants them fresh. Last year we took them to the Amish in Arthur.

It is like a three-and-a-half-hour drive, but they have an immaculate system down there. The birds are very clean, and they are packaged nice. Everything for the farmers' market is frozen."

The Rissmans' innovation and diversity show in marketing and the creation of new items, but this is tempered by a clear realism. "Well, we let our Web site go. For what little sales it brought us, it just wasn't worth the hassle. And we thought about selling our homemade soap, but we found out through our insurance man that there are huge liabilities selling soap over the Internet. We can sell it at the farmers' markets and we are covered under our policy, but the minute we start selling over the Internet it is like a separate business. He said the first time it burns someone's eyes, or it could be a made-up problem, someone out to get you. They could sue you right out of business. So he suggested not to do that. But at the farmers' market we are completely covered. It goes from a farmer's product to being a business product or something like that." Joel and his family are innovative but realistic. "When I was getting my bachelor's degree, I came up with that radio-controlled turn signal. It is something you could mount on the back of a wagon with the controller in the cab that controlled turn signals and brake lights. I looked into manufacturing that, and the insurance was just prohibitive. Because if someone rear-ends you and gets killed, they are going to sue you, whether it was the fault of the blinker or not. So I never ended up doing anything with that."

Joel is always seeking new information and learning from various sources. Joel worked on his master's degree in crop science for several years, driving to an extension of the University of Illinois for classes. "Sometimes it's a video transmission, and other times they just send a teacher out." His organic farming knowledge often contradicted his conventionally trained professors. He described one course: "Plant Diseases – the university has completely missed the boat. I got an A in it, and I was surprised. The first professor was all chemical – live and die chemical. And I thought I basically am going to speak my mind. If it affects my grade, then that is just the way it is. But we were always into it. And there is one example. In class, he says, 'Let's face it, folks. When it comes right down to it, it is yield. Yield, yield, yield.' I raised my hand and said, 'Excuse me. That's not my main concern. My main concern is my bottom line. Yield would probably come third or fourth down on my list. The second thing would be a balance of my soil nutrients. If you have a balance of soil nutrients, the yield will come because of that. The third thing is getting the erosion stopped, getting the microbes back in my soil, and getting all the wildlife and things back into the ecosystem. Once you have those three – the two main soil ones in place

– then that is when your yield comes fourth.' He said, 'Yeah, yeah, yeah. Except for you, everyone else is yield first!' And we went back and forth on that. He knew what he was doing, I will give him that. But he could have tried some other things."

For his oral exam to earn his master's degree, "I just said a prayer, and said God, obviously I can't remember everything, so please only let them ask questions about what I know. So I just took those three courses of the professors that were there and studied their material real hard. So I really hit the nitrogen cycle hard. He just drilled me on it. And I knew pretty much all of it. I looked like a champion. I think he was surprised. And one started asking me about composting, and that was right down my alley because I do composting. And then just a few more questions. So I got the thing. It is not on the wall here, but it is somewhere in the office. But I don't think I am going to bother with the Ph.D. I think I am done. In fact, I know I am done."

Joel is running an experiment on his cattle with various feeds. "I have a grant that I am working on right now with feeding trials with beef. The idea is to have the flavor and taste of grain fed with the low cholesterol of grass fed. And the numbers look very good. Iowa State University is doing the testing for me. I tried at U of I, but that wasn't on their agenda. They haven't changed, nope, not in that respect. I thought it would be nice to work with my alma mater, but they are still wearing their chemical glasses." Joel is keenly aware of the problems of modern industrial agriculture. Describing organic methods, he notes, "Well, we should get more of us doing this, but it is like pulling teeth. My one neighbor said if he doesn't do well this year, he's going under. He won't even consider organic. He won't even say the word *organic*. I don't think I have ever heard him mention it. Hey, take twenty acres and just try it. See what happens. You can always sell the beans on the conventional market if it is a failure. If you are running twelve hundred acres, you can find ten acres somewhere. You won't lose money because you are not putting the expenses into it. That is what people don't realize. You don't have the money for fertility costs and that. You just do a rotation."

The yield versus profit dilemma is clear-cut for Joel. "Everyone around me, I think, is farming over a thousand acres. So one of my goals is to make as much money – net – as the guy farming a thousand acres. I hope I can attain it. I know I will. That is my goal on three hundred acres. And I know that is going to be possible." He is often asked to advise newer organic farmers. "You have got to get the small grains or alfalfa into your rotation. I tell them not to be looking at yield. Look at your bottom line. How much you spent on fertility, how much you worked the ground, and all of those things. In the conventional system, there is such a yield mentality. I can't believe it. You know, 'What did your corn yield?' A few years ago, I started asking them, 'What yield do you want? My *economic* yield?' And they don't bother to pursue it, because they know I will ask them theirs and they don't know it."

"I do know what my economic yield is. There was one time last year or the year before, my uncle was all upset. He saw that the neighbors out-yielded ours by twenty bushel [per acre]. So I said, 'Let's level the playing field.' So I called up the fertilizer dealer and got all of the price quotes. He had a seed plot, so they publish all of the fertilizers and all that. So I priced it all. And then I took out 10 percent in favor of him, fudging the figures in his favor – maybe I made an error in something or maybe he got a discount or something. And that did not include any application costs for chemicals. And when I got it all done, the only thing I didn't figure out was the possible difference in seed costs – he might have gotten a better deal than us or something. And I didn't calculate that in or any machinery costs, just all fertility and herbicide costs. And we ended up out-yielding him economically by twenty-two bushels an acre. And I documented it all right there so he could have it for his files. But that is the same thing – yield, yield, yield is drilled into his head. That is one thing that bothers me about conventional agriculture. They have to start thinking a little differently."

Not that the organic system is perfect. Joel is concerned with recent trends in certification. Many farmers in Illinois changed their certifying agency. Joel explains, "They went to the new one because it is cheaper. Of course they allow you to have GMOs in parallel production. They don't care. And another thing I don't like about them is what they do to get inspectors. They put it up for bids on the Internet and whoever bids the lowest becomes the inspector. You are getting quality work there. I call it the Wal-Marting of certification. You get what you pay for."

In terms of his organic certification, he chooses the highest standards, but Joel has been pragmatic. All his crops are certified, and "Our cattle are certified. Well, the chickens are raised organic, but they are not certified right now because of all of the paperwork. But with the cattle we went ahead with the paperwork and got them certified." The certification issue has not been a problem. "I am getting more and more people to sell to." They sell the livestock locally by word of mouth, plus "our beef sales are now including two restaurants in Chicago. And our chickens will also be in one of the restaurants, maybe both will put them in." In addition, "We sell into the Green City Market, the farmers' market in Chicago. We go there, too." Prices for the direct sales are competitive for consumers and good for

Joel and Adela. "At the farmers' market $3 a pound, whole chicken. But for a whole if it is cut up, we sell it for $3.50. But chicken breast, we sold them for $9 a pound. Eggs we get $3 a dozen. We have eggs, too. We don't have enough layers. We get sixty dozen and sell every egg. Now that is a year-round thing." Plus, Joel sells directly to consumers in the Chicago suburbs. "I started two meat buying groups – fifteen families in Oak Park who e-mail me an order. We put all of their orders together. We drop it all off at a central location in Oak Park. Everyone picks up their order there and writes a personal check. One person puts them all in an envelope and mails them. We do that once a month in Oak Park and in Naperville. We are going to start a third in Wickerville. When they want eggs, it is like thirty or forty dozen at a time."

Joel proudly notes, "I am getting close to 70 percent of our income coming from the consumers." Just to clarify, the 70 percent that is sold directly to consumers is composed of specialty items for the farmers' market that include dog biscuits and homemade soap plus "the beef, chicken, eggs, and turkey. And flax to consumers." Then, with a sigh, "But unfortunately the other 30 percent pays a lot of the bills with the food grade soybeans." So although he and his family are attempting to diversify as much as possible, the reality is that the organic soybean crops are a fairly reliable source of income. But even with the soybeans, they have "taken a hit on that the last couple of years because of quality issues. Well, you know we have had a drought the last couple of years and the bean leaf beetle has been hitting us pretty hard." He describes the pest that is plaguing soybean growers across the region: "What happens is the bean leaf beetle punctures into the pod or into the plant and that allows the vector for the mottled mosaic virus. The bean has what is called a brown swirl, which stains the seed coat. That takes you from a $16 food grade bean to an $8 feed grade bean. So you incur a substantial loss." Japan has been a consistent market for organic soybean exports from the United States, but Joel explains, "What is happening is they are bringing beans in from China and other places. Even the Japanese are trying the Chinese beans because of the bean leaf beetle problem in this area. It is all over Illinois, Iowa, and all throughout the bean-growing region. It is a real problem. The Japanese have actually gone to other sources, even though the taste is not what they want."

Joel describes how he scouts for weeds and pests: "We always walk bean crops. It is not that tough of a job. Basically, I am willing to sacrifice one bushel per acre to weed control. And I'll hire people to walk, but I would spend more than that on chemicals." Then he describes an interesting weeding technique: "I started propane flaming my beans when they are early, and I think that is going to eliminate a lot of hand labor. I've got a machine out there. I spent $1,800 on mine and I will use it year after year. We can go look at it. It is a six-row cultivator, and you take all of the shovels off and put the burners on." That's right, organic farmers use a flame weeder a few weeks after planting their crop. The weeds are burned off, and the crop plant comes back strong. Joel explains, "You flame right over the top of the bean. And it will kill everything in that little band, and then the rest of them – when you start cultivating you want to roll the soil up to the bean so you cover everything else. Yes, the bean itself gets flamed, it does. It singes the cotyledon a little bit, but they keep growing. You really should have the burners in so they were directed straight over the row. It's like a first flush. The little grass was just two leaves. You flame it and kill it. I am convinced that if you can burn off all of the weeds that come up with the beans, I can control everything else with a rotary hoe and cultivating."

As is typical in organic farming, information must be sought from across the country. When Joel initially bought his flame weeder, he "consulted with a man in Minnesota, and he says they've been doing it up until the first trifoliate. He's been off of chemicals since 1972 and organic since 1985. He said he'll even flame his corn when it is clean, because he has always gotten at least a four-bushel increase in yield. He says it is generally a four- to eleven-bushel increase. So he said even when his corn is clean, he will flame it anyway, just for the yield increase. Just having the flame on the corn causes a yield increase. And these are side-by-side trials, and he doesn't know why. No one has been able to explain it. Maybe it is the shot of carbon dioxide from the flame or maybe the stress changes something in the plant. I don't know. But he says every year he takes side-by-side tests." Many organic farmers conduct their own experiments and disseminate these true research results among their peers.

Although Joel has a background in agricultural sciences, reads extensively, and conducts his own on-farm research, he still jokes about some organic techniques that he has learned. "Everything that they say about conventional farming doesn't apply to organics. They always tell you the earlier you get your stuff in, the better, and I shot myself in the foot this year. I got out there too early with the planter after two years of wet weather and being delayed. I wish I would have waited until about the third week in May. I told Adela, 'Next year, tie me up, lock the shed, cover my eyes, plug my ears so I can't hear or see the neighbors, or maybe just go on vacation and come back the first week of May.' The old saying is that if you can sit bare-assed in the field comfortably, then you can plant. It's the truth. This is scientific organic here! If you can sit there comfortably for five minutes, then go ahead and plant. But when the ground is cold, you want that seed, as soon as it starts imbibing water – you know, expanding – you don't want that process to stop until that thing is out of the ground and growing." Proper planting techniques have been discovered by trial and error.

Pest management is evolving on Rissman's farm. Foxtail and velvet weed are no longer a problem. "My weeds had shifted to lambs quarter and pigweed and those are fertility weeds. So I really don't have grass problems anymore. Just rotary hoeing and cultivating. I might try some biodynamic spray. It's not much of a problem. Last year it was tough because we got a four-inch rain right after we planted; I mean the night after I planted. So I couldn't get in there to rotary hoe for two weeks. It wasn't that bad. The weeds were there right with the crop. They didn't come up afterwards. Which means they matured with the crop. Everything was dried up when the beans were ready. So we didn't have to wait for a killing frost or anything like that." Joel's pragmatic approach to weeds is one that is based on his experience: a few weeds don't really cause any problem, as long as they can be easily sorted out at harvest time.

Joel has not had to buy much machinery on his own. "I am buying my dad's equipment on lease agreement. Now the repairs are starting to come. Everything is twenty or thirty years old." But he does all the repairs himself. In addition, he uses his own seeds from year to year. "All of my small grains and flax and one variety of soybean I will use." The exception is when there is a problem. "I am undecided this year. If my wheat kills off, I might have to buy a spring wheat. And depending on the market, which I don't have any contracts yet for the beans, will determine if I have to buy seed. One variety I couldn't use because of the brown swirl. I have seen it bleed over into the next generation. I have proven that to myself, and another farmer has proven the same thing."

They have no permanent hired workers on the farm. "But we are looking to get an apprentice or something like that. We hire temporary help. Transient workers just for a couple of weeks. That is about it. Last year we had an apprentice and right up at the last minute he told us that he only wanted to work one weekend in July and one weekend in August and which weekends would I want him to come? I told him to forget it. He wasn't going to learn anything in a weekend, after he said he was going to work the whole summer for us. We have gone to the organicvolunteers.com [Web site] to see what we can come up with. Maybe one person. Other than that it is just Adela, me, and the kids. That is a pretty heavy workload."

Through their hard work and innovation, the Rissmans have a successful farm, one that is constantly evolving. Joel is pleased with his decision to shift completely away from growing corn. "I think my sorghum yield is 140 bushels for the farm average. My neighbor who farms across the fence says his farm average for 1,200 to 1,500 acres was 118 of corn. Yes, and my seed cost is only about $3 per acre versus his $20–30 for hybrid corn. I planted it June 25. So virtually all of your weeds are done. I cultivated it once. I didn't rotary hoe it. I cultivated it once and did nothing more until I harvested and then I combined it." So, for his operation, the sorghum is providing a more economical and better livestock feed crop. There are differences as far as the animals, too. "Well, any hybridized crop, where man has tinkered, has a negative energy to it. Because grain sorghum, sure they call it a hybrid, the male and female parts are not separate on the plants. It is like hybridizing wheat. It has a positive energy. The chickens are supposed to do better on less. I am finding that with the cattle. I feed very little, and it has really put some pounds on."

Joel firmly believes that the quality of organic food has positive benefits for people's health. "We have local customers that come here to buy. We have had I don't know how many people, giving testimonies in the yard that they hadn't started healing until they started eating organic. And the latest person, he went to the Mayo Clinic in Minneapolis, and I forget what rare type of cancer he had, but they just threw their hands up. They didn't know what to do with him anymore. They basically said, 'Go home and get your house ready for you to die.' The first day he came here with his wife and he couldn't even get out of the van because when he would stand up his blood pressure would skyrocket. He went on a diet that is all organic. Two months later he can walk forty-five minutes a day. He is really starting to look good now." According to Joel this is not an isolated incident; rather, there are "tons of stories like that where people are coming in here that are our customers now. It is amazing. It makes you feel good. We did more good than the doctor."

Guided by Joel's constant experimenting and desire to learn, this farm is highly diversified and evolving into even more markets. "With a lot of my small grains – we have a pretty good sideline doing custom chicken mixes for small farms and large ones. We have a guy who has about a thousand laying hen, organic. He gets feed that we custom grind for him. That is something that we accidentally fell into." And the farmers' market is completely new for Joel and his family. "We just started that last year. We sell our beef, chicken. And Adela's homemade soap – all organic and we use our organic tallow, too, so there are no toxins. And then all organic oils and essential oils for the fragrance. There are no colorings in it. If the FDA has to approve it, we don't want it in our soap. Now we are also going to sell wheat grass kits. So we will sell the soil mix and the grains and people can go home and raise it themselves. And we have a new product in the market, Kendyl and Noelina's Canine Crackers. We are making our own dog biscuits for the farmers' market. We have vegetarian, beef, and chicken." They sell at the Chicago Green City Market at Lincoln Park every Wednesday from May until October. "It is about an hour and fifteen minute drive with no traffic. At five in the morning, you can get there in that time. But it is fun because not only are you a salesman, but you are an educator. There are a lot of people – I give my sales pitch for flax feeding of livestock – and now I will have the research information of my own to back it up."

HOME ON THE PLAINS

Allen, Cliff, and Naioma Benson, Colorado

This large dryland organic farm lies on the open plains of eastern Colorado. The landscape is wide open and the winds are truly awesome. The horizon seems endless. It reminds a visitor how small a human truly is. This is the land of the buffalo, the native tribes, then the settlers, and of course the Dust Bowl. This is where Naioma and Cliff Benson have been farming organically since 1977, and they've now handed most of the work over to their son Allen, who is in his forties. This makes for the fourth generation on the land. Cliff's grandfather homesteaded here in 1909, and various parcels of the farm were inherited from both sides of the family. Allen gives a few details about their farm. "We're at 5,700 acres right now, both rented and owned, 4,800 in crops. We're spread out over a nineteen-mile range, it's nine miles by ten miles," a patchwork. Naioma explains, "This is real advantageous, too, because of the hailstorms. We won't lose everything at once."

The shift between generations has been a gentle one, as Cliff jokingly asks his son, "Allen, what year did you come back and start working on the farm? I know when he showed up, I started to slack off." Naioma says, "You must have bought your first land in 1996." Cliff explains their labor: "Now it is pretty much me and you and another hired hand. Summertime we hire another man sometimes, too." Allen clarifies, "The second one is part time."

Cliff and Naioma talk proudly of how they made the transition to organic farming. They were the only ones in the area. She says, "In the seventies a stranger was visiting here from the east coast, and he stopped and told us to check into organic farming, because he saw a few weeds in our fields. The weeds meant that we were already not spraying much if at all." Their land

Cliff, Allen, and Naioma Benson at home in eastern Colorado. (Credit: author)

has low rolling hills with clay loam tops. They had stopped using chemicals because they were ineffective. They did not increase yield enough to offset the cost. "It doesn't make sense if your input costs are higher than the production increase in earnings." In addition to this economic reasoning, Cliff describes the typical mind-set: "I ran into a guy the other day who said, 'How in the world do you control your weedy season?' I said, 'Well, I just use crop rotation,' and he says, 'We don't think we can raise a crop out there unless we spray it two or three times.' It's just the mentality. The chemical companies have instilled into them that it's the only way to go." Allen adds, "Chase down a gallon of killer and take care of the weeds."

Naioma elaborates on the work involved in their crop rotations: "Our neighbor, a fellow older than Cliff and I, he thinks it's pretty crazy to do all this work with the spring crops. He says, 'God made this land for wheat and fallow, wheat and fallow. We should not be doing any rotation here.' But at his age the work is too much for him anyway." She goes on to describe all the work that their crop rotation involves. "It's tough. Allen is going all summer to fall. He's planting millet, buckwheat, hay millet all in June, plus trying to get combine equipment ready for July. He finishes harvest in July,

Allen Benson is also at home in his fields. (Credit: author)

and it's time to get the seed beds ready for wheat, and at the same time the millet, buckwheat and everything else is coming off, so it is triple the work. No doubt about it."

Allen says, "I'm doing the right thing because I don't like playing with chemicals. I know some people hurt from the anhydrous," referring to anhydrous ammonia, a common volatile fertilizer used in conventional farming. Naioma reflects on family memories. "Do you remember years, and years, and years ago when we used to use the chemical Thiamet? And L. J. used to put it on the little boxes on the back of planters. Just handling it and pouring it in there out of the bag. The smell of it would dilate the blood vessels in his nose so much that he would start gushing blood. Oh, it would scare me to death." Allen adds, "Well, there was one guy who got it on his fingers and didn't think nothing of it; smoked a cigarette and he got sick. It put him in the hospital." Cliff brings us up to date: "Other folks still use it around here. Well, top dressing and starter on their seed." Allen describes other chemicals used by conventional farmers in the area: "They are wicking their rye with Roundup all the time." In contrast, Cliff sums up the advantages they experience with organic methods: "Yeah, you don't have to be afraid of crawling in a bin to clean it out and getting knocked down by a chemical that's been left in there. It's good without that stuff around."

Since they are organic farmers, they are not experts in chemical use, as Naioma explains: "When you get out of the mode of chemicals, and we've been out since the seventies, you are just out. We got a survey from a chemical company, something like – do you recognize our product? What's the label color for Roundup? It went on and on, a real intense survey. And I thought, boy, have you hit the wrong people!" Allen agrees. "Well, the telemarketers call and try to sell you a chemical, too. You tell them you are organic and, click, they hang right up." "And all the money the chemical companies spend on advertisements!" Naioma exclaims.

The Bensons are outside the conventional system and have developed a complex farming operation with crop rotations, on-farm cleaning, loading of grain, and extra "custom" work that Allen does for other farmers. First, in terms of their crops, Allen explains, "Mother Nature does the best thing for the rotation – controls it. I don't know what the rotation will be. It depends on the year and if I have wild grass [weeds]. I just play it by ear." Cliff says they do whole-year rotations rather than multiple crops within the same year. This is due to the fragile ecology of this region; little rain, high winds, short growing season, and thin soils add up. Allen describes a few of their crops: "Buckwheat is a seventy-day crop; millet is ninety." Cliff always has a light-hearted comment to describe a rough reality. "You stub your toe, why, then you wait until next year. That's why they call it 'next year country.'"

Their crops are unusual for the area and ever changing. Cliff fondly describes one crop: "Have you seen buckwheat? Black like a little coal and then when it starts blooming in the summertime the whole field looks like it'd snowed. It turns white." Allen adds, "Looks like it snowed in July." Naioma says, "We were the first to plant buckwheat in Colorado. Cliff and I went clear to Dakota to get the seed and so then didn't know how to handle it and had to do a lot of calling around." Naioma turned to her husband. "I think you talked to someone in Tennessee on how and when to harvest." Cliff says, "I talked to several people around the United States, trying to figure out how to handle it." Bringing the conversation back to the present, Naioma reiterates, "But it does bloom a beautiful white. It stays white for ages. We have people pull in the yard, and they stop and jerk out a plant and ask us, 'What is this stuff?' Any more of those people and we won't have anything left of it!" Allen turns to the practical side of the buckwheat crop. "It's an excellent green manure," meaning it can be plowed under to fertilize the soil. Cliff says, "Well, they said it was real fertile for the ground; builds fertility in the ground. So when it volunteers the next year, we let it grow up fairly big before we go out and turn it under. That's just reseeding because the [harvest] combines are throwing it over your way. And that's a great fertilizer. You turn it under the next year. So you get the crop plus the fertility the next year."

Allen adds, "We've been raising buckwheat, wheat, millet, and alfalfa. And hay millet; we are going to try to get it in this year, too. Hay millet for cattle." Cliff describes how they can earn more by doing most of the finishing work on farm, as opposed to many conventional growers who deal with more "middlemen" because they simply "dump" their grain in an elevator. "A lot of our crops are shipped clear to Europe, but we clean it and seal it and do it all here before it ships out." Allen adds, "Yeah, with a spray cleaner." Naioma describes the equipment. "That is a new innovation. We bought parts of it in Canada." Cliff goes on, "And then our container loaders stick them in containers." Naioma says, "That's a piece of equipment Allen and Cliff invented." Allen says, "We rebuilt the cleaner, too, because they didn't have it set up right for our sized grains. We ran around here for six months until we finally got it right."

Naioma sums up the economic aspects of this work: "So I think that has made a difference with our sales. Allen is cleaning the millet and then using this loader to shoot it into the containers and fill the bulk head, and then it's sealed." Allen says, "It's like a Salad Shooter to load the containers. We want to patent it." Naioma takes us a bit farther in the process: "It goes from here onto the truck, then railed from Denver to the East." Allen says, "And it depends. It goes to either the Canadian port or the Houston port." Allen describes some of the paperwork associated with exporting organic crops: "We have a broker, but we still have to do the shipper's declaration and all that stuff. Every commodity has its own special serial number, so they know what it is with the serial number off the shipper's declaration." Cliff is glad to have handed this chore to his son. "I'm glad I don't have to toy with that anymore."

But simply listing the crops they grow does not paint a full picture. There are multiple varieties, plus specialty crops and seeds grown, and each sold to numerous grain brokers or other distributors. Allen downplays it. "Mostly we're doing hard red winter wheat." Naioma elaborates. "You do two or three varieties, though." And Allen admits the complexity of his work. "Three varieties and I sell it to different people. Sometimes people just want the one special variety." Naioma prods, "Now you're in there planting a new variety." Allen says, "Yeah, a couple hundred acres of hard white winter wheat. It's high protein, and it goes to Denver." Naioma notes, "The seed is incredibly expensive. And that's a patented seed, so he has to sign a contract and he has to account for every bushel." The markets vary from year to year, and Allen says, "A lot of our wheat is going to brokers out of California. All brokers are kind of quiet." Naioma continues, "This broker on the west coast has not been willing to share with you whether it is being exported or where it is going. But you're not dealing with him this year, and your buckwheat will be going back to Europe."

With a bit of a grin, Cliff describes another new marketing channel they've been having some success with: "They're buying wheat, some of them now, ya know, for organic juices." Allen explains, "Ya, I got a buyer who is buying it and juicing it. It's big in California. I just sell them the seed, then they grow it in flats." Naioma continues, "Those natural foods stores usually have little flats of wheat grass growing, and they just snip it off with scissors." Cliff says, "Well, they grow it, you know, for the juicing. They want a certain variety because it grows much faster than the other varieties. It takes off. Some wheat comes up and just lays there. Others are real vigorous and just come right up. That's the varieties they want." Finally, the grins turn to real jokes, as these true wheat farmers find it amusing to think of drinking wheat. Cliff says with a twinkle in his eye, "They say a little shot of vodka goes pretty good with it!" And Naioma jokes, "An all grain drink. Well, you know how interesting Boulder is, and that's where all the juice bars are. The guy who is juicing it has a big market throughout Boulder, and he says there are some real bars that serve it so you can have the green grass juice and a shot of vodka." Allen is more realistic. "I think that alcohol would kill the protein. So much for it being good for you." Joking aside, this conversation illustrates that the Bensons produce high-quality organic crops that are sought after by specialty buyers, and they are clearly proactive in seeking new marketing opportunities.

Naioma explains other diversifications. Allen does custom work for other farmers. "Allen has had to develop a manure-hauling service. He has geared up the equipment to do that. Cliff started by buying one spreader; to spread on our ground and what do you have now, two spreaders?" Allen replies, "Well, one and a half, because I didn't want to have two complete headaches. I went in with another guy who has a feedlot." Naioma says, "But then you have a belly dump [truck] that you haul." Allen explains, "If it's a long haul. I do it for other farmers. Conventional." Naioma says, "But you usually work two loads for you, one load for me. Or for cash, right?" Just another way to add to the farm income, to diversify and to stay in business.

Naioma explains how she built up their marketing opportunities. "The way I pushed our marketing was by going to the 1992 Natural Food Show in Anaheim, California. I went and checked labels. Made contacts. Gave out my business cards. Now I really want to go to a German show. You have to find a market and then grow the right crops. But you need to show a good product, reliable crops." So it is not just the contacts that are important but the Bensons' quality crops. Yet Naioma cautions, "It is very, very competitive. And you'll do a disservice to people if you write a book and tell them they can go out and find all the markets in the world tomorrow. I believe it was fourteen years before we ever sold a bushel of organic wheat, something like that. Couldn't find a market for it."

Organic certification is critical for the Bensons, since their marketing opportunities, especially exports, depend on it. They had certified with the Colorado Department of Agriculture, but it is on hold until the federal regulations are in place. The Bensons were both positive and negative on state certification. First, "the inspector was just wonderful to work with," according to Naioma, someone who could give them information about weed issues. But the state certification was not accepted internationally. "It just never got the recognition, and we tried and tried. We'd ask, 'Why don't you use Colorado certification?' It was better for us because we don't have to pay an assessment on it and it is cheaper. And mostly we didn't have to wait and wait like with the other certifiers." Cliff notes the certification quality was high. "It was a good certification, too. There were a lot more stipulations than anything else." Naioma elaborates, "What was great about Colorado certification was that if you sold a product that was not certified organic it was a felony charge. And we'd tell them that, but it was still never recognized in Europe. But we kept paying up because that was the way the rules were written. And we had a good inspector. He always had good ideas when he came and we learned a lot. We dropped them this year, and we dropped the other one, too." Allen shrugs. "Long story."

They had some frustrating experiences with one certification agency. Allen says, "You got the local crew and the inspection committee and you go through all the reports and send it all to the main headquarters. They'd have to go through it a second time. They don't take the farmer's word for it, and then your inspection certificate shows up late. Well, here you are trying to move your product, and they want a current certificate. Where is it?" Naioma says, "The year before we pulled out, we lost business because it took us eleven months to get a certificate." Allen says, "We had everything done and inspected by the end of October, and it still didn't come out until February, and we took care of all the rough cuts and did all of the paperwork." After all this frustration, it was time for a change. They are now certified organic through another agency, Quality Assurance International, which has a joint agreement with Bio-Swiss, a certification favored in Europe. This helps them market their grain in that region.

The certification process can be onerous, particularly to farmers who tend to be rather private people anyway. Allen says, "All of the neighbors ask me about organics, but then when they see the stack of paper you gotta do, boy, that quits right there. And then, dealing with brokers. Other farmers are used to taking their grain to an elevator and if they want a check right away they can get it. Well, you don't do that when you deal with a broker. He's gonna play with your money for ninety days. If you're lucky." Cliff agrees. "Knock on wood, we've never lost a load or anything to a broker. They always come through, but sometimes you have to be a little cranky with them. 'The check needs to be here.' Of course we never let them get in arrears very far. About two loads, and that's as much as they'll get before we get a check." Allen explains, "Some are from overseas, the rest are back east, so you have to wait for them to send it to you. You got to take into consideration I put it into containers, then it goes to Denver to be put on the rails, and then the rail might take two weeks to get there, then it might have to sit in the ship yard three or four days, then it's put on a ship. So that's five weeks to two months." The Bensons' grain is geographically well distributed. But getting their payments sometimes takes longer than the slow barge of grain. Allen tells of one situation: "Well, luckily I called the shipping company, and they knew when that ship was supposed to be there. So I called the broker and he said, 'I don't know anything about it.' Then he called me back the next day and said, 'You were right. It's in here.' So I said, 'Are you going to cut me a check this week?' Well, he wouldn't call me back." It took several more months before Allen got the payment.

In addition to specific farming concerns, the Bensons are aware of the influence that agriculture has on their community. Rural change is obvious in the community of Sterling, an agricultural town of four thousand that's about twenty minutes away. As the ag economy is in a shambles, towns have had to seek other economic options. Allen explains, "The prison went up about three years ago." Naioma shakes her head. "Sterling thought it would bring an influx of money. It didn't. Well, our hospital lost a lot of its nurses. They went to the prison because they are state employees and they have every benefit on the planet and tremendous job security. A lot of the prison guards were transferred in. They are a really cliquey group of people. They don't participate in town. They don't do volunteer work. They aren't a churchgoing group." Cliff says, "I was a little concerned when they brought that in. Not that the prisoners would escape, but, you know, who follows them – the type of people they buddy up to, but they are controlling that real well. They got dogs sniffing the cars out there, and the prisoners have to tell them who is visiting and they do a background test. The sheriff will stop them right there if they have a warrant against them." Allen cracks a grin. "Yeah, they caught a few like that." Naioma raises an eyebrow. "If you've got a warrant for your arrest, why would you go visit someone in jail?"

Then the conversation returns to agriculture, as is often the case with this family. Naioma says, "The prison here has made a big demand for little ranchettes or farmettes, whatever you want to call them. They want forty acres and a house." Cliff describes land prices: "I keep thinking it's gonna go down with reduced commodity prices, but it is staying right about $400 an acre. Allen agrees. "Yep, $400 to $410." Naioma: "That little house out there along the road was my aunt's, and she just sold that last week to someone at the prison. That seems to be the only people buying any acreage." Times are tough for conventional farmers.

Naioma reminisces, "Life on the farm is different from the next generation and ours. I got to be at home, and Cindy [Allen's wife] has to work because they need her health insurance benefits. What is it now, is it 60 percent of farm men have off-farm jobs now? And for women it is more like 80 percent. I haven't seen the numbers for a while. Last I saw it, it was staggering. Years ago, I knew how many women were working off-farm, but I was surprised to see all of the men." She knows the issues well and has clearly thought about the decline of U.S. agriculture. She sees the effects of agribusiness in her region. "It's terrible concentration. Vertical integration – Cargill bought one-fourth of Colorado's grain elevators. They have train cars, mills, seed. They control exports. They do black-balling: 'Use our type of seed, not that other one.'" Farmers don't have much power within this system unless they seek other options as the Bensons have. Naioma says that the only way to slow this dreadful trend is "by not being held hostage in a wheat-fallow rotation. Increase diversity! Maybe go organic."

At the same time, Naioma's opinions on environmentalists were clearly guarded. "There are too many radicals. What is the goal? Do they want continued production or to eliminate U.S. agricultural production so we need imports? We need a balance. Then they are helpful! In the eastern United States, the topsoil loss and groundwater contamination had to stop, but it wasn't their livelihood that was threatened." At the same time, she hopes that agriculturalists are also environmentalists. "Well, the majority are. Full-time farmers have to focus on future generations in heart and mind until bankers push them the other way!"

Economic woes are often joined with climatic hazards on the dry plains.

Cliff notes, "We average fourteen to sixteen inches of rain." Naioma weighs in. "No, twelve to fourteen inches." Allen sarcastically notes, "Hey, lately it's been one and a half to two inches! To be truthful, we've only had two and a half inches since December, and it's the end of June now." Naioma nods. "Last year was dry, too." Cliff has a half smile. "That's hurting the guys worse that are nonorganic because they got a chemical bill on top of it all." Naioma agrees. "And they put that chemical down on their wheat last fall before they knew the summer was going to be this yucky."

There is a complete lack of understanding about organic farming in the region. They laugh as they describe one recent event. Naioma begins, "Something happened, the container left here went to Denver, put it on a flatbed rail car and the rail was to run from Denver to go back out east. They got to Akron, Colorado, and the train wrecked. So here's this load of organic millet lying there on the ground. So one of the guys called." Allen says, "They wanted to know what 'organic' means." Naioma continues, "The rail company is looking for somebody to salvage this and pay them to haul it off. He asked, 'What can I expect to get for organic grain?' Well, you can't sell it as organic. Once the seal is broken, it is no longer organic! They said, 'Oh, it looks just fine.'" Allen gave more details. "That one guy said, 'Well, it is still organic,' and I said, 'Did it spill on the ground?' 'Well yeah.' I said, 'Is that certified organic ground?'" Naioma said, "And the rail companies are famous for spraying the right-of-way." Allen finished that story shaking his head. "Then the one guy called asking for our certificate copy, so he could sell it organically. I don't think so! No way."

Bensons have a clear and positive view of their future in organic farming that may even involve another generation. Allen says, "I would like my kids to stay in farming. I don't know if they will stick with it. I hope so." He speculates, "The future of organic ag? That's tough, I don't know how it's going to end up. There are different aspects opening up. It's going to hang in there, but I don't really look for it to increase greatly, do you?" Then Allen thinks about this issue of food safety and says, "One thing that might make it go is if for some reason the cattle market has a disease problem. One of my millet buyers sells our product in forty-eight countries. When mad cow disease went through England, their organic sales went up 30 percent." Naioma is surprised. "Oh wow, I hadn't heard that." Allen emphasizes, "Yup, 30 percent because of mad cow disease." Naioma digs into this issue. "And they are only selling grains and nuts and stuff like that. They are not selling meat." Allen concurs. "They shied away from the meat after mad cow." Naioma considers. "That's interesting. Thirty percent. That is a variable, isn't it, if we have a major food crisis in this country?" Allen agrees that organic farming would "really boom if something like that ever happened here." This part of the interview was conducted in 2002, so Allen had some premonition of the U.S. cattle situation. In addition, Naioma's opinion is that "the next bump in the road will be with the federal regs. Who is going to adapt to that, and how?"

The conversation turns back toward the family, since the generations are now making a transition in farm management. Naioma asks her son, "What changes are you going to make?" Allen says, "I don't know. Everything is so spur of the moment. You know me. Something will pop in my head, and . . ." Naioma smiles. "The thing about youth is they have no fear of trying anything. We had to wade through it and think." Cliff says, "A little luck can make things work, too." Naioma adds, "If you have a little rain, it helps."

Driving around the fields, Allen proudly points out their beautiful stand of wheat. And Naioma warns, "Organic farmers are innovators. If America loses family farmers, they've lost innovation." She should know. She and Cliff were true innovators when they went organic in the late 1970s. Allen is following in their footsteps, developing new equipment, techniques, and markets.

MARKETING ORGANIC VEGETABLES

Phil Foster, California

Phil Foster is now about fifty years old and operates a 250-acre certified organic farm in the San Juan Valley of northern California, about two hours south of San Francisco, but he started out as a city kid. "I had family that farmed. Not immediate family, but relatives in this area. I was just starting college and I started working with them in the summers, and that is when I became interested and decided I wanted to go into agriculture. At that time I was going to UC Berkeley, and then I switched to UC Davis, got into the ag program and got a degree in ag science and management." When he got out of college he worked as a farm manager on a huge conventional farm, until he settled back in this area and worked with his cousin on his conventional walnuts, apricots, and cattle ranch.

Then, in 1987, Phil rented seventy acres and started growing tomatoes for a local cannery. "It just so happened that the neighbor across the road was a small-scale organic farmer. So I got a chance to visit with him. In hindsight, I didn't really think too much of the way he was farming. It seemed kind of sloppy, not very well organized, but at least it opened my

Phil Foster on his California vegetable farm. (Credit: author)

eyes up to something different." He learned that there was a small organic marketing company looking for more growers, so he took five acres and started growing about ten organic crops. Times were tough. "It was very hard to make money on cannery tomatoes, especially on a small-scale farm. Seventy acres was a small amount for California. I just spun my wheels. There was no money made. There weren't really any good options." But his organic acres did turn a profit. "The five acres that were organic, we made a little money off that. And it peaked my interest in soil fertility, how to grow crops without commercial fertilizers. All these things that were kind of new and interesting." Since his initial organic certification in 1989, his farm continued to evolve. Now he owns 30 acres and leases 220 more. "It is in two different locations that are approximately twelve miles apart, but in that twelve miles, there is quite a significant climate change, especially in the summer." This is beneficial, as "we can grow lettuces, celery, and kind of cool season crops in the summer here, and then have some crops that are relatively warm season like melons and watermelons, which are usually not grown much in this area."

The three-year transition to organic certification is often a major stumbling block for farmers converting their conventional ground. Phil was lucky

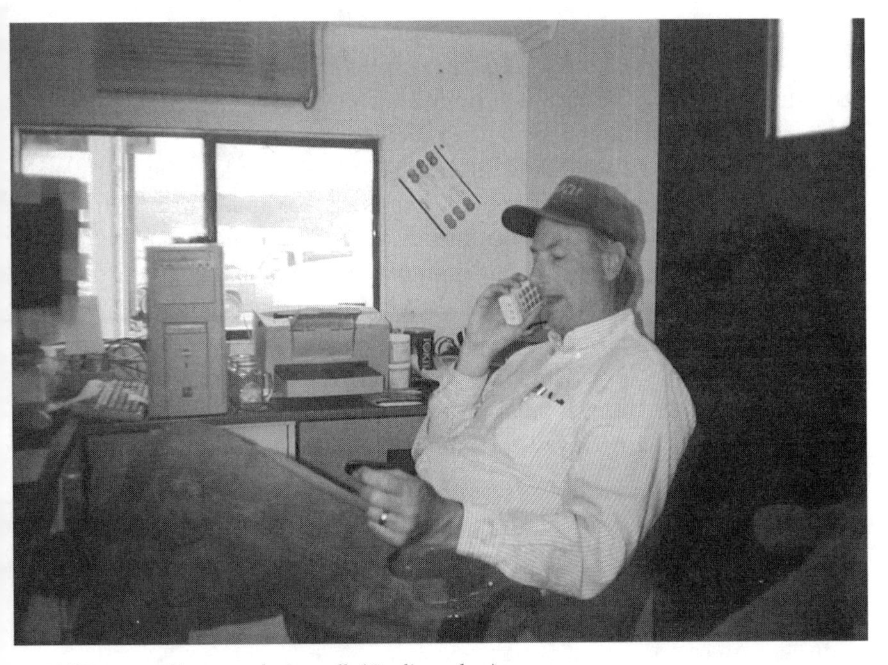

Phil Foster making a marketing call. (Credit: author)

in that some of his land had not been conventionally farmed previously. "So we started everything in the process, and we had some parcels of land that could go into certification right away and some that had to transition from previous history. The fifty acres of tomatoes that I had farmed conventionally – and actually toward the end of that tomato crop, I started managing it and we did our last aphid spray with soap. So our last prohibited material was much earlier in the season. This thirty acres here was planned for a subdivision, and they had put in cherries five years earlier. Well, the subdivision never materialized, so it was an abandoned situation and we were able to transition this piece into organics, certified organic the first year. And then we rented twenty acres next to us, and we had to go through a three-year transition on that parcel. We are farming on a total 250 acres now."

As in many parts of the western United States, water availability is an issue. Here, the farm relies on well water, and it is sufficient. "Since we have small blocks, and do succession plantings, and utilize the winter for crop production, we can crop year-round with different crops. Maybe some of it is fallow on certain years; it just depends on what is going on. We are on drip irrigation, so we can make the water last longer. We watch how much we have going on. We can sprinkle at night, drip during the day. The orchard is on minisprinklers, so it is not so critical on irrigation timing as, say, a vegetable crop." But water conservation is definitely a concern. "Especially during the summer we have to be careful on how much we are doing because we can only cover so much with the water we have."

Land prices vary depending on the quality and exact location, but even agricultural land sells for $15,000 to $20,000 an acre in this area. "What we are faced with here is a lot of influence from development. We are close to San Jose. We do have a moratorium in our county that will slow things down a little bit, but I do think the long-term speculation on land value is really pushing land costs up. On rent, the twenty acres that we rent here we pay $600 an acre. Which for our area is a good fair rent, maybe a little on the high side, but it is adjacent to our thirty acres, so it is well worth it. The other ranch we pay $200 an acre. It is not in a farming area. It is more of cattle rangeland area. It is good bottom ground, it is good soil, but it isn't a ranch with a lot of water." These numbers seem phenomenally high, especially to those of us not accustomed to California land prices, and it makes one wonder how organic farmers can survive in this climate of high land costs and low crop prices. The answer is diversity – an incredible diversity of crops.

"We have six acres of walnuts, three acres of cherries (and this is not going to be exact as we go along because I won't remember them all exactly). Peppers are a big part of our operation. We do a lot of them wholesale, and we are mostly doing colored peppers, red and yellow. This is a good area for growing colored peppers because it is warm enough where we can get good production, but it is not that hot where we have to worry about sunburn problems. We are doing about fifteen to eighteen acres of peppers a year. Those are at staggered plantings, so we have peppers from August until Thanksgiving. It just depends when the frost hits us and knocks us out. Onions are key to our operation. We are on a five-year rotation on alliums. So onions, garlic, shallots, and leeks will adhere to a five-year rotation. I'm not too sure about the leeks, but the dried alliums will stay on a five-year rotation or as close to that as we can. We do between garlic and onions and shallots, about thirty-two to thirty-five acres a year. We do a few acres of early onions to get our season started. Those are generally sweet onions that don't store that well. The bigger acreage we usually do twenty-two to twenty five of reds, yellows, and whites. The yellow and red on an equal percentage and the whites a smaller percentage. Those onions we can store. They are varieties we can go into cold storage with them or store on the ranch up until Christmas. Then the ones in cold storage we can come out with until April. And come out with good quality and not a lot of loss. Garlic we are probably at six acres a year. One acre of early garlic and five or six acres of late garlic. We are getting to where that six or seven acres will be entirely local. Maybe 20 percent of it will go wholesale. And shallots, we are only doing something less than an acre, and it is all for local. That is the right amount of shallots to keep us in shallots through the winter."

He pauses for a breath and continues. "Lettuces, we are probably, by the time the year is over, between thirty-five and forty acres of lettuce. But we are doing bigger wholesale type plantings. In the spring we have plantings of five or six different varieties of leaf lettuces. We are doing three- to four-acre plantings. Then once we get through the end of May, when a lot of the other companies come out with lettuce, we drop out and only keep lettuce plantings in that are half-acre plantings to meet our local needs for the rest of the year, through the summer – up until Thanksgiving. Sometimes we might have fennel or spinach that goes in, maybe a few rows of spinach. They get moved around different acres of the ranch. If we do twenty to twenty-four acres of lettuce wholesale in the springtime, they will be in different areas of the ranch and the last planting will come over to this ranch. And then we are doing all our summer plantings on this cooler ranch.

"Cabbage has been pretty important. It is nice to have brassicas in your rotation. We are slowly transitioning into more broccoli. We probably do between red and green cabbage, napa, and bok choy, probably fifteen to twenty acres a year. And half is in the spring and the other half starts up harvest around Thanksgiving or in November and goes through the winter. That is one of the crops that we can be pretty successful with in winter harvest. This winter was the best cabbage market I have seen ever. So you just never know what is going to be good and what is going to be a dog. We have had some pretty doggy cabbage markets, but we have gotten to where we can maintain some pretty consistent yields. Between broccoli and cauliflower we probably do ten acres a year, really more so on broccoli. And broccoli has a really nice rotation. It is something that we try to do small plantings throughout the year. It works well in our local route. It is just a nice thing to offer. The problem is I don't know if we ever make money on it. We make money at times. Since we are in it all the time there are some windows where the prices go up. Do we really make money after the year is over? I don't know. But it is one that is essential to have. It is a nice item to have on your price list. People will usually buy broccoli."

Phil pauses. "Now I have to stop and think. We do beets, probably a couple of acres a year. I might be surprised. Maybe it is three or four acres. We do red and gold beets all for bunching. Then again I am not sure how well we are doing on those kinds of items. When the market is good, we can do okay. There is a lot of hand labor involved. It is one of those crops that is like a stepchild. We have put multiple plantings in very small scale. Do we always watch them as well as we should? Do we always keep them as weeded as we should? Are we hitting the water just right? Are we that efficient with harvesting compared to someone on a larger scale?" He is almost apologetic that he cannot be as efficient as other farms. But he concedes, "Certainly we can make it work locally because it is another item to add to the list of things we offer.

"We are doing bunch carrots, which is different in this day and age, when carrots are so efficiently grown by the regular seller carrots. We are doing a specialty variety carrot. We bunch it. I bet we will end up doing four or five acres this year. But that is maybe one-third of an acre being planted every two weeks for three-fourths of the year. So we have carrots nine months of the year. That is one we really need to work on. We need some time to figure out if we are doing this as efficiently as we should be or if we are just spinning our wheels. We are doing a little bit of parsnips, probably less than an acre a year. We are doing less than an acre of rutabagas. We are probably doing an acre of leeks, and that is one, every now and then, even something you do on a small scale, it is surprising how efficiently and how well received it is from our buyers. It seems like we can do a good job on leeks. Whatever we are doing on it. We haven't grown them that long. It's not like we are experienced growing them. It is either the right area or we just happen to do the things right, we can manage it fairly easily. So leeks are one, even on a very small scale, there is a pretty high profit margin. And it is a nice one to have in our rotation. Fennel is another one. We probably do two acres of fennel a year and have it over a lot of the season. Even when the prices are low on fennel, we can still make a little bit of money because we have such a high pack out per acre. Some things I grow because I like eating them, and some I have no idea why I grow because I have no desire to eat them. Fennel is one of those crops."

He thinks a bit and continues. "We have chard. That is one we just started up doing, and it looks like another one of those crops on a small scale we can make it work fairly well. It is something we do in the winter. I would say we might be running an acre of chard this year, maybe more. We are probably going to try some at other times of the year. Some of the things we do are dictated by pest control. We do sweet corn; we probably do about twelve to fourteen acres a year. Most of that is going wholesale, and it is going wholesale for the month of July. We try to hit the Fourth of July. We do eight hundred boxes a week wholesale for five weeks. The reason we do that five-week period is because historically that is an in-between period for corn earworm. There is just not a lot of pressure around for the month of July. It is in between flights; it is before the generations build up. We can be down to as low as 0 percent worms, and not go above 15 percent or 20 percent worm damage. And the worm damage is just tip damage. So 20 percent damage . . . which for organic is very acceptable." Indeed, that is very acceptable even in conventional chemical farming. Phil explains, "Corn is just a nice thing to grow. People like corn. It works well in farmers' markets. It works well with our local route. There is something to be said about a nice ear of sweet corn even if you cut the tip off if it has good flavor.

"And we grow cucumbers. We usually end up doing about an acre and a half a year, and that is mainly for our local routes. We are still trying to learn how to do cucumbers. I don't know that we have ever made money on cucumbers. If we could do a good job with it, local cucumbers throughout the summer are a nice item to have. And then in the fall sometimes the prices go up. We are doing summer squash. We are doing maybe four acres a year. Maybe five. Two to three goes to wholesale, and then after that there are some smaller plantings to spread it out through the season. We still do a wholesale summer squash deal. We try to hit what is an early market for around here in May and June and then try to get out of it before the local people come in because that usually means the prices go down."

While planning and growing these diverse crops takes a lot of time and commitment, perhaps even more effort goes into marketing the many types of produce. Up until two years ago, Phil sold only to wholesalers that distributed his products regionally and nationally. "They had a sales staff. They oversaw the cooling and shipping. That worked very good for a number of years. We were able to expand our business as the organic industry grew. There was a big increase in growth from 1990 to 1998. So we grew our business over those years. We were able to have a pretty diverse crop mix rotation." But then things started to change in organic production. "It looked like it was becoming a little more competitive. We were trying to figure out where we fit into the whole thing. We have a pretty good medium-sized farm for organic production, but there were more people getting involved, there were bigger growers, conventional farmers trying organic in this area. And what was happening was they weren't experienced with organics, especially marketing and size of fields. They would plant somewhat the way they were used to planting, just much too large. It ended up there were pretty depressed prices on certain things. And we thought there would be more of that. As you got more people into that and organics became a little more mainstream, it may parallel the way the conventional market is and that, to me, is not a healthy way to market. The farmer is more at risk."

So Phil hired Terence Welch, who had experience as a partner in an organic foods wholesale business, to develop his marketing strategy. Although Terence is active in multiple aspects of the farm, his knowledge of the organic food market systems has really propelled the farm in recent years. Trying to avoid the price vulnerability found in the wholesale organic markets, they "talked about setting up a local delivery system. We started delivering directly to stores in the Bay area and Santa Cruz." But this required a financial commitment, too. "In order to get set up, we had to invest in a small cooler on the ranch. We put in a seven hundred square foot cooler. Since then we have added a couple smaller coolers to give different temperature rooms for the different products." They also do some packaging on the farm. "We have a packing facility for our peppers. A small packing line. It is a packing line that I bought used from someone ten years ago, and they probably used it for twenty or thirty years. We have modified it a little, but for our size it works pretty well. We also have a very nice onion grader. It is a field onion grader where you pull it through a field. It makes our onion packing very efficient." They also have two refrigerated trucks, an 18-footer and a 22-footer. "During the busy season we are delivering three days a week and we are doing our own sales. Terence was doing it on his own, and since we have gotten busier we have hired another person to help with sales. From essentially two or three years ago, we have gone from something like 95 percent of what we sold was going into the wholesale market to what is projected this year 40 percent wholesale."

The remaining 60 percent of Phil's total sales are a complex assortment of clients. There are smaller retail stores. (They are small health-food stores that didn't always get good service in the wholesale distribution system. They are happy Phil's farm can "work with them directly and give them good service.") Also, a retail distribution center that supplies a large natural foods grocery store chain. And a restaurant and a caterer (they are near the grocery stores, so it is "easy to drop off, it isn't much more time"). Plus a couple of processing companies (his red peppers are ingredients in organic soups, for example). And also three farmers' markets. Phil's wife is a high school science teacher, "so she has summers off. Her involvement with the farm has been mainly with the farmers' markets. Eight years ago, she started doing some farmers' markets in the summer." Now the farm sells at these farmers' markets year-round, and this forms an important 5 percent or so of total farm sales.

He describes some of the current trends in organic production, with more agribusiness interests and larger farms going organic. "There are some organic farms with several thousand acres. I know one conventional operation that has maybe six thousand acres of organic production. Crop rotation: I think that is the dilemma. They are carrot people, and they need that ground for their carrots. They want a certain volume of carrots. This is separate from their conventional carrots, which is thousands of acres. In their organic they may have six thousand acres, but they are really looking at two thousand acres of carrots they can do and it is a three-year rotation. They have these other acres they have to figure out what to do with. So they are trying to get other crops in. I don't think they want too many crops to make it complicated, but they have to find some other complimentary crops."

He explains the farming concerns. "As far as the larger scale, there are a lot of problems, I think, with trying to grow organics on a large scale. It is very management intensive. You can cookbook conventional; you can work out a system. You know what you are going to do at certain times of the year. If you have a pest problem, you know what you can do for it. You can manage the fertility quickly if you need to adjust something. Well, organic is much more difficult. If you get to a certain time of the season and your plants are running out of steam, there is not a lot you can do quickly, or the things you can do are very, very costly. I think they are going to find that the systems aren't in place and they will be vulnerable in certain ways to pest problems. I don't think it lends itself to organic as it does to conventional, where you can really manipulate more things."

This agribusiness presence in organic farming has other impacts, as consolidation of processing and distribution hurt smaller farms. Phil worries that the processors may shove him aside. "In the future are we going to be so small that people don't want to mess with us? They basically run our garlic in a day. I don't know that we are going to invest in the specific packing equipment for garlic, so I don't know what we will do in the future, but I would like to keep garlic in our rotation as much as we can."

Phil explains, "Pinnacle is our brand name. We have that labeled on our boxes. That is the other thing; we are maybe a little different now. We tried to promote our label early on. We didn't pack under someone else's label. That is a lot harder to do now because there are not as many people marketing, and they want you to market as a certain label. So if a grower is growing for one big label, they are not using their own label. They are packing in an agribusiness owned box probably." But Phil is trying to keep his own identity and remain independent from this system, although "those opportunities are much more difficult these days. There is a lot more consolidation. Our future in wholesale, I think we will always have product, but it may only be five key items that we can do particularly well, like onions. We have always been able to grow onions well. I think we could always be fairly competitive on onions." But as agribusiness becomes increasingly involved in organic production and sales, the smaller, diversified farms may find it difficult to compete.

Phil and his employees work hard to maintain this complex and multifaceted cropping and marketing system. "My wife works about a quarter of the time. Terence is full time on a salary, and Robin, our second salesperson, works year-round, but is not full time year-round. One employee that has been with me since the late eighties is now on salary, Efrain. He is pretty much our foreman for the main two hundred acre ranch. He lives on the ranch. He is integral in our whole operation. He is really a key employee. It varies a little, winter to winter, but we usually have all the full-time work for the winter either live here year-round or want to stay here for the winter. That could be anywhere from twelve to eighteen people. And our field crew for the busy season probably goes up to thirty people, and even at that we have a two-month period in which we utilize a labor contractor and have another eight to ten people who end up working full time for those few months. Those months are August, September, and October. Our wage rate is $8.25 an hour up to $10 an hour [in 2004]. So we have really tried to keep our wages as competitive as we can. The job on a farm is a hard job, working ten hours a day, six days a week, sometimes a little bit longer. At least we have diverse jobs. It is not like someone is hoeing weeds all day long. Somebody might be picking one type of crop for a few hours and then switch to another crop. They may be helping irrigation for a few hours; they may be weeding for a few hours. We have jobs in the shop packing and then tractor driving jobs. At least the work varies. It is not so tedious. Even though it is hard work, there is variety to it. And then the men that work on the ranch, a lot of them know each other; they have been working with each other for a while. So there is a good working relationship among the men on the ranch."

Phil feels that a great deal of his success is because of his employees. "We really have a core of people who have worked steady for us for a number of years, so they know all the jobs. Then over the years as we have needed people we have added more people and those have become more permanent employees. Maybe they go to Mexico for three months in the winter, but their jobs are waiting for them when they get back. And we have been able to pay . . . well, the people that work on the farm aren't compensated nearly enough for what they do, but it's as well as we can do. We have been able to slowly increase their hourly wage. We try to give cash bonuses. We have a health insurance plan for the workers and their families if they live here. And the fact that the farm is organic may help, too. "I think the workers know a little bit more. I think the people are attracted to the higher wage. People are always coming and asking about work. I am usually pretty cautious about hiring people. I usually let Efrain make the decision. I talk to him and he selects people when we add to our workforce."

The paperwork necessary for certification can be time-consuming. "It is a fair amount of work. Certainly in an operation like ours, where we need to keep track of all the inputs. We do have that computerized. I am keeping track of all that material; anything that is done on a daily basis and put it into a field journal. Then Robin will take that field journal, and the ones that are inputs, like seed, or transplants, fertility inputs, compost. . . . She takes those journal entries and puts them into the computer so we have, or we can make, an on-farm input report easily now. So at least that helps. I have other people helping me now. I used to do all that myself, and it was quite a bit of work. Terence has a good record system for tracking all of our sales. All of our invoices have all the organic information: grown in accordance, CCOF certified. So we are covering a lot of those bases usually the inspection goes rather smoothly. We are improving our paperwork all the time. It takes a lot of effort, there is a lot of work, but it seems like we have a fairly good handle on that now."

The farm is certified to both the California organic standard, and the international IFOAM standard, which is not really necessary. Terence explains that the real value of the international certification is that "it really helps when you want to sell out of the country and since we don't really do that, IFOAM certification doesn't really benefit us financially. But Phil is liking to be on the cutting edge of certification. He likes to be one cut higher than what is required of him. So I think for him it is more an ideological thing as opposed to a financial reason to grow to IFOAM standards."

Phil has been active in the California certification agency, which has helped him meet other organic growers. "Well, I go to various meetings with CCOF and there is a chance to interact with other growers. I am on the certification standards committee, I used to be on the board of directors. So I got to visit people from all over California, because it is a chapter-based organization. So I got to meet a lot of people over the years, rub elbows with them."

It is good to meet with other organic growers since they can exchange information. Phil says it is difficult to get good information on organic farming, and when he started he "read all those books out there. Books that people had written for organic farming. The Eliot Coleman book on organic gardening, I got a lot of information out of there that I could use. A lot of that had to do with soil fertility. Whether you are small scale or large scale, it works on both ends." For example, "We do make our own on-farm compost. We make two thousand tons of compost a year. We have a compost turner. We bring in dairy manure, and we bring in hay or straw. I used to grow my own hay for that, but I found it was cheaper to buy that. We also get the clean green material [yard waste like leaves and grass clippings] out of San Jose. There is a big push to keep good green material from going into a landfill. We do get that material and use it. I would say we are probably bringing in 75 percent of our [compost] feedstock between the dairy manure and the clean green material." Of course, this sounds like organic farming on a large scale, with such a volume of manure, but in fact, "I would say we're medium-sized, but we are farming it more like small-scale farm because of the cropping, smaller plantings, the succession plantings. It is not different from a large scale, but we are growing fifty crops, so it is more like what a market gardener would do or a small-scale person that has a more diverse cropping system.

Although Phil started out in conventional agriculture, he wouldn't go back. "I have learned too much about organics. I've bought into it. The pesticide deal, I would not like to get into again. I've done all of that. When I was the agronomist for the large company, there were some years that were particularly bad pest years. We would have the airplanes out spraying one or two thousand acres a night, and we would do it four or five nights a week. Now I have just seen how things can work a little differently. I don't think I would ever be, if I were a conventional farmer, a large-scale farmer. I like farming for my own. I think part of the problem with conventional is that you are being pushed to getting bigger in equipment and acreage. I think you really put yourself at risk doing that. Not that we are so stable. We could have things happen to us where we may not stay in business. But I think we have a little better of a chance. I think I was lucky at the time and got into organic at the right time, and that may be the only reason I am in business for myself now. It is hard to say what would have happened. I, for sure, won't ever go back into conventional farming. I am going to make this work, and if it doesn't, well then I'll work for an organic farmer or find something that I can do."

He sums up his farming experience: "I think in general the farm has been pretty good to us, and it has given me the type of employment I have always wanted. So there are some other benefits there that are hard to put your finger on. I want to continue farming for myself for another thirty years. I don't see any reason I would retire. I am doing something that I like every day, something that is 90 percent incredibly enjoyable. I have always liked farming, I have always liked equipment, I have always liked growing things, and I found out that I like growing a lot of different things. I like diversity. There is a lot to pique your interest and keep you sharp. I work with some really good people. I like the area here. We have been able to at least own a small portion of our farm. We have our house on our farm. Those may seem like really simple things, but it is something I really enjoy. I like walking out of my door in the morning and being at work."

5

Making It Work

The present scientific quest for odorless hog manure should give us sufficient proof that the specialist is no longer with us. Even now, after centuries of reductionist propaganda, the world is still intricate and vast, as dark as it is light, a place of mystery, where we cannot do one thing without doing many things, or put two things together without putting many things together.

– Wendell Berry, "*In Distrust of Movements*," speech to the Tri-State Environmental Educators Workshop in Evansville, Indiana, October 1998

Farmers teach us a great deal about the complexity of organic farming in the United States. Steve, Mary and Rob, Joel, Phil, and Cliff, Naioma, and Allen represent various geographic regions, manage very different types of farms, and describe things in distinctive ways that relate to their individual farms. Organic farmers are not all alike, but their common experiences converge along several themes: economics, ecology, society, and personality. Within these broad categories are specific traits and influences that are often present among successful organic farmers (table 1). These influences come to life through quotes from the five farmers we know. They describe how things actually work on their farms, which provides a rich understanding of the geography of organic farming throughout the United States. This helps us "map" organic agriculture and chart a path for its future.

ECONOMIC FACTORS

Organic farmers most often point to the influence of economic factors in determining the success of their farms. Clearly they must achieve economic sustainability in order to remain in farming at all. The leading economic

Table 1. How It Works: Important Influences

ECONOMIC	ECOLOGY	SOCIETY	PERSONAL
Markets	***Balance***	***American Culture***	***Independence***
Diversification	Ecosystems	Conventional ag	Information
Direct marketing	Rotation	Views of organic	Risk
Instability	Antipesticides	farming	Low debt
Big Organic	***Weather***	Cheap food	Busy
Crops	***Soil Health***	Rural regions	***Innovation***
Quality	***Science***	***Policies/Information***	Experiment
Organic food prices	***Environmentalists***	USDA standards	Learning
Organic Opportunity		Certification	***Tradition***
		Research	Evolving

component is marketing. Where they sell the crops and livestock they raise is the most important aspect of their success. Also important is the high quality of their crops. Organic farming has provided them with an economic opportunity that might not have been attained through other agricultural systems.

Markets

A successful organic marketing system takes time, skill, and constant effort. First, the sales must be diversified. Second, direct marketing can be a key approach for some farms. Third, there are concerns about organic market instability. Fourth, farmers are worried about the future, as large agribusinesses exert increasing influence in the organic marketplace.

Diversification

Cropping and marketing diversity are the roots of success for organic farmers. Phil, in California, explains it this way: "We don't make big money on any one crop. We might hit some good markets on specific crops and have a very good year with specific crops, but we don't have a lot of crops that do too terribly bad. So we have quite a bit of enterprise. We have the orchard, we have the delivering, we have the sales, diverse crops, and we are in it year-round. So we are not where someone has hold of our money before it gets to us. Certainly someone we sell to could be a bad customer and not pay us, but that would only be one of many customers."

For citrus, diversification means marketing. Mary explains how smaller citrus groves in Florida simply cannot compete. To stay competitive, it's necessary to market independently. "Most of the big packinghouses will not pack for you. They will buy your crop if it is quality or if they think they can make some money, but they are not just going to pack it and charge you. They want to make money off of you. People get so greedy. We have a packing facility, and I have a citrus fruit dealer's license. So I can go out and buy and sell with that citrus fruit dealer's license."

Speaking of the organic grains they grow in eastern Colorado, Naioma says, "You have to find a market and then grow the right crops." In addition to the six organic grains that the Bensons grow and market (with multiple varieties that often go to different buyers), Allen has also developed a "custom" operation in which he does specific jobs for other farmers in the region. Of course, the Benson family has been growing grains for generations, but they are always willing to tap into a new specialty market. The latest is selling wheat for juicing.

In Illinois, Joel says, "I am diversifying as quickly as I can here. The finances show that." He describes how his family now sells beef, chicken, turkey, eggs, and other items locally to restaurants, through buying clubs, and at a large Chicago farmers' market. He is trying to reach every possible customer. "We have our organic soap, and now our dog biscuits, and barley grass and wheat grass sets, so we can get some of the vegetarians' money, too. The problem when you sell meat is the vegetarians pass you by." But this is no longer the case, thanks to the Rissmans' diversification.

In upstate New York, Steve notes that their original livestock farm has changed a great deal in recent years. "Growing the vegetables has been a challenge, but it has been, for the most part, profitable, which helps. We saw vegetables as a way to expand the farm, saleswise." The fact that vegetables are highly perishable means that harvesting and transporting must be accomplished quickly. "The vegetables need to get out the door and down the road." He notes, pragmatically, "Being diversified is great, but it can kill you when you are trying to get everything done."

Direct Marketing

By taking out or reducing the "middleman" – the buyers, wholesalers, distributors, transporters, processors, and stores – organic farmers are able to keep more of the profit on their farm. Terence, the California organic marketing manager, says idealistically, "I always tell people: the most revolutionary act you can commit is to go to a farmers' market and buy from an organic grower. Because then you have bypassed the whole distribution system. You're buying food that's local, so you're supporting your community;

you're supporting an agriculture that's benefiting the earth. It's amazing, you know?" But Phil, the owner of this California vegetable farm, is more practical. "That's a much smaller scale. And farmers' markets would be very difficult to make your living in. You have to do all the growing and then you have to do all the selling. Our experience with it is you really earn your money at them."

Yet, in the face of increasing competition from agribusiness interests in organic vegetable production, Phil has indeed shifted his marketing emphasis for the better. "The big-scale organic farms have certainly made me plan my marketing, and it is one of the forces that influenced me to change our marketing strategy. As an aside, probably what I really wanted to do was market more locally anyway. I thought it was ridiculous. Here is a very good area for organic produce, certainly Santa Cruz is, and I think San Jose and the Bay area have a lot of potential. Here we are shipping wholesale, trying to ship to these other regions of the country. The local people have to go through the normal distribution channel. From broker to distributor to maybe another middleman before it gets to the stores. I just wanted to see more of my produce staying local. And I thought I could supply a good quality produce by going direct."

Speaking of his CSA in upstate New York, Steve says, "We are working to improve it. I feel there are certain crops people really want to see in their bags each week: some type of salad greens. Having salad greens in the bag each week, or a little better variety each week, and such. There are things we could be doing, but my brother and I are just so swamped with the things we need to do, and we just don't get to it." He hopes to put more effort into this direct marketing in the future. "The CSA is not done yet."

For their citrus, Rob and Mary have decided that direct marketing is not worth the added time for the marginal profit increase. Rob says, "There are advantages and disadvantages to it." And Mary notes, "It's feast or famine" in terms of customers buying varying amounts of products at different times. For them, the packinghouse and wholesaling have proven to be more stable and profitable. Likewise, for grain production on the Colorado plains, the Bensons most often work with brokers who deal nationally or even internationally with organic grains: millet, buckwheat, and wheat.

Defying the odds of big Corn Belt agriculture, Joel's farm in Illinois has recently shifted from selling through brokers to more direct sales. He says that financially, "It is tough right now, only due to the drought. The direct marketing is kind of holding us up. If I were a crop guy, I would be done." Although the farmers' market is a new venture, "It is fun. We begrudgingly did it out of desperation, financially. That's how we got into the farmers' market. I always thought the farmers' markets were a big pain, but I don't think that anymore. It is fun and good income. Going directly to the consumer with chickens and beef – that's marketing. All of those extra dollars in processing go in your pocket. You cut out everyone else, and that is how it should be, if you ask me. I am anti–big establishment."

Instability

Overall, though, there tend to be problems with organic markets because they are relatively new and thus somewhat unstable. Naioma describes selling their Colorado grains: "A major downside to organic is dealing with exporters. There is a need for bonding of export buying agents. Since it is a rapidly growing industry, the downside is the flaky people you have to deal with." Indeed, it took them eight months to get $60,000 from one buyer. Another time they had to hire an attorney to get the money they were promised because a buyer arbitrarily decided to pay a lower price than they had originally agreed upon.

Since organic markets are growing rapidly, some wholesalers and distributors see them as a good way to make some fast money, which can be problematic for farmers. In California, Phil explains, "There have been some stressful times, too, that have probably been very uncomfortable. Sometimes when the money came in slow, some people that we marketed through went out of business and we had to wait for money from them." By doing their own local sales, they have "solved some of those things," but "there are still a lot of stresses that go along with it."

Mary, in Florida, sums up the pros and the cons to organic marketing: "The reason I do this work is because I love dealing with the people I deal with. You can do a handshake deal with them. Their words go. But I dealt with one person and their word was not good. They call me every year, and I will not deal with them. I think I made myself clear that they will never load at my place again."

But Joel describes the situation as improving, at least in terms of some grain markets. "There is a lot better market now for wheat. Back in the 1990s you were limited to one or two people, but now with the cracker and bread market, pastry dough, pizza dough, tortillas, more and more organic products are in that area, so there is a greater market for wheat." As these markets strengthen, organic farmers will be less vulnerable to unreliable brokers and wholesalers. At the same time, a stronger market may pave the way for larger agribusiness interests to expand within organic production and sales.

Big Organic

As organic farming and the popularity of organic food continue to increase, agribusiness interest in organics is also on the rise. Phil describes the situation in California: "Maybe you don't see that in the Midwest, but certainly here you have some pretty significant conventional produce growers doing organic. Or a portion of their operation is organic. One of the big lettuce people, they are very interested in organic. You have a lot of people taking a look at it." In terms of the California example, Phil explains what is happening in organic production: "Well, that is the negative thing that is happening in the organic market. It used to be that a small grower had the opportunity to market product wholesale. Some of them with five or ten acres, those opportunities are becoming much more limited, unless they have a specialty crop, like kiwis. There are still opportunities, but there are less and less. In the early 1990s there were a number of produce brokers that were like grower agents. These grower agents wanted a lot of people growing for them. So there were opportunities for people with twenty to fifty acres to market that way, whereas now those types aren't around anymore. They have either gone out of business or been bought out. And the choice for people to market to is less and less. You have to have something that they want, and you have to have enough of it. Because now you are a vendor, and they probably don't want to deal with someone with a small amount."

To support some of the smaller-scale buyers, Mary describes her sales of organic citrus: "A lot of times, if the distributor is so high, then the little guy can't buy it and he goes out of business. The distributors are buying and reselling to the small guys, because the smaller guys can't take ten or fifteen cases at a time. So they have to go through a distributor. I try to work with them. If I know it's going to a small store, well, I'll try to take off a few dollars." In addition, Mary says it is unfair how the larger producers are treating smaller growers. "It is not right what they do. They are putting farmers out of business. The farmers aren't making money. I am only because I pack it and market my own. We are making money and putting it back in here and building the facility." Rob is worried that "in ten to twenty years, if more people keep coming on board, the organic market is going to be just like the conventional market. They are going to ruin it."

Joel says he's heard that agribusiness interests are getting into organic policy. "As far as the organic program, now the Georgia contingent tried to slip in a thing on the farm bill saying they don't have to feed 100 percent organic. I don't know who these Georgia people are, but they are even paying off the politicians down there and getting what they want. Isn't that a sham? We called our legislators and talked to other people that had and they said they were getting a lot of calls on it. I am hoping they get more and more. If they are going to do that, then the organic community ought to unite and let the word get out on the people who are not feeding 100 percent, and the consumers can boycott it." There are many avenues for agribusiness to make inroads into organics; farmers and consumers must act to block each one.

Crops

Two economic aspects of crop production are often discussed by these farmers: they produce high quality products, and they depend on organic price premiums. These are both crucial to their economic success.

Quality

Organic crops of today are a far cry from the small, wormy veggies of days long past. Now farmers are proud of, and depend on, the high quality of their crops and livestock for maintaining their sales. According to them, it is not enough to simply produce by organic methods. Rather, their success is based on the fact that their products are excellent. Regarding their Florida citrus, Mary states this clearly: "We're marketed as superior quality, and I'm very picky about my quality. And I don't worry about competition. I'm not going to worry about the guy next door, because if you sell quality you will stay in business for a long time. It's that simple." Rob proudly notes that major natural food stores across the United States stock their citrus. "Mary stopped in one of the stores in Atlanta and met one of the Big Shots. She said as soon as you walked in the store, there were our Orlando tangelos. They said they were the best-looking pieces of fruit in the whole store, and they were right when you walk in the doors. The produce guy said, 'Nothing beats these things – for looks or taste.' "

Because he is not one to brag, it was hard to pull a statement from Phil on the quality of his vegetables. In fact, it is a given; otherwise, they wouldn't sell. Finally, after much conversation, he mentioned his recent addition of apples to their diverse California organic farm: "We have some red delicious. And everyone moans and groans about red delicious. A lot of times they are picked early so they can get on the market, or they are picked to store so they aren't picked with high sugar content. But if you let a red delicious get a nice sugar content to it and sell it that way, they are a really nice apple."

For their New York CSA, Steve can judge quality by the returning members year after year. He stays current with the members' needs through feedback from a year-end letter and a survey. This helps him decide what crops are best and what changes need to be made. For example, he has decided to grow a more constant supply of lettuce and to expand into some crops that he would not otherwise grow. This seems to please his CSA members.

In Colorado, the Bensons proudly display their certificate, as Cliff reads it: "100 percent pure organic millet." Allen explains, "I do have a purity test done every time for every container. Kansas Grain out of Paxton does that for me." Cliff explains, "Purity means no weeds or any other foreign material." Speaking of organic markets in general, Joel explains, "There are hundreds of people out there for buyers and brokers. But I'll have to say it is a quality market. It has always been a quality market. So nothing in that respect has changed." Just as crop prices that organic farmers earn vary, so do food prices that consumers pay for organic foods.

Organic Food Prices

Growing food organically costs more, since there is no simple short-term chemical "fix" to rely on. Thus the price premiums earned by farmers are passed along to consumers. Phil, in California, describes the situation with onions and garlic: "They are much more difficult to grow organically than conventionally. The prices are definitely higher than conventional, and I think they should be for a couple different reasons. If you look at the cost and all the things of somebody that is really doing a good job organically – building the soil. There is a lot more cost going into farming organically. There is certainly more labor involved with weeding. And I think it is nice to know that there is something out there holding a higher price. I hope that organic continues to do that."

In terms of his California produce, Phil describes the organic price markups: "When I used to figure in my mind wholesale, if we were getting X for it, it was selling for twice that to two and a half times that in the store. But there are brokers, there is trucking, it is a lot more. I don't know what the rule of thumb is from when we deliver to the stores. I think they mark up at least 50 percent. But we are getting the sales, and we are getting the trucking as we are delivering directly to the stores. They may try to mark it up double sometimes, but I think it generally is a 50 percent markup. If we are getting a dollar a bunch for broccoli, then it is probably selling for $1.40 to $1.60."

So the markup on organic food can be substantial, particularly as wholesalers and distributors realize that consumers are willing to pay more. Steve, in New York, explains that it does cost more for organic food transportation because it is at a smaller scale. "You have to truck it, and maybe one or two other people mark it up. You take it to a wholesaler, and they mark it up and deliver it to a store, and they want to make some money on it. And you are not moving a lot of volume, so that adds up. It is not as efficient as, say, the conventional produce system because you are not moving as much volume."

Mary realizes that their prices are high, but it is linked to the quality. "I certainly don't expect people to buy it at that price if it isn't what I think it should be." So organic prices should be closely linked to quality. Rob says, "You ain't never had lemons like these. Believe it or not, the general public don't hardly know what lemons like that are. They are used to those little, dinky, dry-ass California lemons. In Florida, the lemons are the best. Do you know what those things sell for on the market? Why, $29.50 for a carton. They are worth their weight in gold. Conventionally they sell $20–22, same lemon."

Finally, based on his Illinois experience, Joel discredits the common economic assumption about organic production. "Why is the price going up if more and more people are getting into it? We haven't reached that point of supply and demand yet. In my opinion, there are more and more people every day who are saying they have cancer in their family or something, and they are looking for alternatives. Organics is what – maybe 1 percent consumptionwise? And over a year, they figure it is up 13 percent. So we have at least 12 percent of the population to go before that supply and demand situation will kick in. So what is the population? Two hundred and sixty million. So you figure 12 percent of that. It's going to be awhile, I think. I doubt that price is just going to take a nose dive."

Organic Opportunity

Organic agriculture can provide an opportunity that conventional agriculture does not. According to these farmers, profits are higher with their diverse organic farms than if they had conventional operations like their neighbors. In Colorado, the neighbors joke about all their new techniques, but the Bensons are leaders. "Neighbors see the good financial situation with organic; we're buying land, updating. You can't argue with success," says Naioma.

Joel also started out as a conventional farmer in Illinois corn, but realized "the chemicals weren't working. I couldn't make a living. You struggle to pay the bills, and we had no insurance, nothing. Living on a major highway, and my equipment wasn't insured. We couldn't afford it. We just couldn't make a living." After converting to organic in 1996, he says, "Now, we have some money and we are getting insurance for our things."

In Florida, Rob bluntly notes, "One thing about farmers, most of them are pretty damn stupid. They are used to doing things one way and they can't change to save their life. But anybody that has enough balls to change their operation and they are losing money and can't make it . . . The ones that change over. I tell them if you can't put it on the organic market, you can put it on the other. You have the best of both worlds. Whereas with what you are doing now, you know what you got."

Speaking of his family's upstate New York farm, Steve says, "We were in a financially competitive position here, and we couldn't get big enough to survive by just getting more ground and better equipment, getting bigger to survive in that way. With 100 acres of vegetables in the conventional market, I am nobody. With 100 acres of vegetables in organic, I am not somebody huge, but I am still a player at certain times of the year. It gives us a little leverage that way." Organic farming has provided an opportunity to survive in agriculture, which probably would not have been possible with conventional methods. In addition to economic factors, ecology plays a crucial role in the success of organic farms.

ECOLOGICAL CONCERNS

Organic farmers are comfortable with ecological conversations – not necessarily ones based on strict scientific definitions (although some would enjoy that, too) but rather a discussion of how things really work on their farm. The key ecological factors, according to these farmers, are soil health, weather and climate conditions, and ecological balance. They also have interesting perspectives on environmentalists.

Soil Health

Soil building is the basis of production success in organic farming. Phil, in California, says that when he first attempted organic farming, "it piqued my interest in soil fertility, how to grow crops without commercial fertilizers." It seems that this interest in the soil is what motivates many organic farmers in their decision to continue with organic methods. Joel says his Illinois farm has seven types of soil. He recommends an important step in the organic farming process: "Start out with soil samples. And don't have someone come in and do it. Take the sample yourself. Then you know how hard you have to push down on the probe, and over the years you can say, hey, this is getting easier. Because you are not killing off your microbes anymore with chemicals, and the soil has a chance to loosen up. And you are more in touch with what is going on. You can observe different things – why are these weeds here and not over here? But you will see the weed patterns change as the years go by. I have seen it."

On their Florida grove, Mary says, "To properly maintain your trees and have a quality product, you need the mineralization, you need a cover crop for the nitrogen, and then some people think doing nothing is organic." Rob says, "We have the best dirt in the state for citrus. They call it the Fruit Land Peninsula." He says that his methods ensure that he'll have "nice good-looking dirt" year after year. Allen describes some of his family's Colorado farm: "This is good ground here." His father, Cliff, says, "Buckwheat builds fertility in the ground."

In Illinois, Joel adds compost to increase his soil's fertility. "I do all of my own composting. I compost all my manures, except the soupy stuff if we get a lot of rain. But I generally combine it with high Calcium Lime and then sheet compost it for the winter." One of the college courses he took "really helped me understand what is going on. Before, I knew maybe half; I didn't really know why. So now I understand why." Joel, like most organic farmers, relies on compost to increase soil fertility. So in terms of costs, "You figure that I have spent so much less on inputs, because my whole fertility program is my compost and high Calcium Lime." In California, too, Phil says, "We make our own on-farm compost. We make two thousand tons of compost a year. We have a compost turner." Steve says that for his New York crops, soil fertility is achieved with cover crop. "Our biggest one is red clover and all of our small grains. We seed it basically when we plant barley in the spring. Or if we planted wheat the fall before, we'll frost seed some clover."

Weather

Weather and climate conditions play a crucial role in the management of organic farms. For Florida citrus, Rob explains the impact of freezing temperatures: "We could lose everything we've got in one night. We were right on the borderline a week ago. I've got all early fruit here, and when I start the month of December, I don't care if it takes twenty-four hours a day, I'm going to get it off those trees and sell it as fast and as hard as I can. If I run myself into the ground like I am right now, so be it, because it takes one night and you are out of business. I would rather know I have the bread in hand than have it sitting out here frozen on a tree, not worth a nickel and watch it drop to the ground."

In Colorado, where precipitation is only thirteen inches a year, and May hailstorms are common, Allen says simply, "Mother Nature decides what we plant." In Illinois, precipitation is usually abundant, but not this year; Joel says that his winter wheat may not survive due to a lack of snow that provides insulation. "It is kind of looking like mine might have winter killed because there is no snow cover. Well, I can try some spring wheat, hard red spring wheat. I'll try that."

Organic farmers note that the specific climate conditions of their farms are beneficial. Phil notes that parts of his land vary. "The easternmost ranch can get really warm, whereas this ranch has a real influence from the Monterrey Bay in keeping it quite a bit cooler, maybe as much as fifteen degrees on a lot of days." Weather is a key influence on farm decision making, and organic farmers have an excellent knowledge about climate patterns on their farms and in their region.

Balance

Conversations with organic farmers often touch on the topic of ecological balance. In addition to this notion of balance, they also discuss what constitutes an ecosystem and describe the importance of crop rotations in maintaining soil fertility. They further note how pesticides disrupt agricultural ecosystems.

Organic farmers see complex ecological relationships and realize that these influence their crop production. Rob, in Florida, observes, "Well, if you have a good organic program, it will withstand drought and it will withstand extremes a little better than the conventional stuff." Based on his Illinois experience, Joel explains the very premise of soil balance, as opposed to the conventional, chemical mentality: "I read in this chemical magazine once, this professor said, 'Even one weed is too much.' He didn't say anything about trying to get a mineral balance in the soil, trying to increase the microlife that will make these seeds not want to germinate. Weeds are growing there to correct a condition in the soil that we, as farmers, have screwed up. When that condition is corrected, they won't grow there anymore. I have seen that on my own farm here. They don't once address that problem. They just say that one weed is too bad. They've missed the boat." Phil notes that on his California farm, "things that we find work on the ranch have to do with more than just applying beneficial insects or having a hedgerow. It has to do with the fertility of the soil, how we are building the soil. It has to do with the native populations of beneficial insects. The research is that ladybugs are not that good of a beneficial insect to release. We are having some success doing that. Is it because we are doing something different? Is it because we are not using any pesticides?"

Organic farmers love their land, and they have an affinity for regional ecology. For all his tough talk and swearing about Florida citrus, Rob hushed as he pointed out an eagle soaring above us. "Look above you. We don't call it Eagles' Nest Grove for nothing. This is the largest concentration of nesting bald eagles in the southeastern United States. They don't bother nothing."

Ecosystems

Farmers are not apt to use fancy words for things that seem obvious, and organic farmers are no different. Cliff may be hesitant to use scientific terms, but his ecological knowledge about the Colorado plains is apparent. "I guess you call it 'the ecosystem,' don't you? And I think with chemicals you upset it so bad because when you use the chemical it kills the good bugs along with the bad bugs. So then you have nothing to combat one another. Organic is better. You leave the ecosystem alone, in balance. Don't mess it up."

Joel also expresses complex ideas about ecology, based on his Illinois experience. He says that once you get off chemical dependence, you "let the natural flora and all these insects that attack the pests – predator insects – all come back. You know, chemicals can't distinguish between good and bad and they will kill everything. That is the sad story. But once the predator insect is let into the ecosystem again, it bounces right back, too."

Rotation

Related to their ecological understanding of balance and the importance of soil health, organic farmers describe how they use crop rotations to build their soil and prevent pest outbreaks. In Colorado, Naioma notes that their crop rotations are complex and determined by many factors, such as weather or the appearance of specific weeds. "If wild rye contamination is bad (from neighbors or from blowing or from some other combine), then we'll plant millet there, because it kills out the rye and the jointed goat grass. This drives inspectors crazy! You just can't set a clear pattern in this dry land. Our rotation indicates that you need a hands-on operation. Somewhere else maybe you can plan A, B, C, D, then E crops. But here, you can't. What if a field gets hailed out in our May hailstorms? Then we have to look and see. We'll pull the wheat out and put millet in."

Joel described some carefully planned intercropping that has worked well on his Illinois farm: "What I'll do is with my last cultivation, I will mark this seeder on the back of the cultivator in the center. It has a little electric motor on it. And I interseed my corn with red clover and vetch and a little bit of rye. That way, you take the corn off and the other stuff will take over. It covers your ground all winter, and it helps erosion. Then in the spring, I will probably take my stalk chopper and chop it up and plow it under. Then you've got the greens, your nitrogen source, to go with the stalks, which is your carbon source. If you plow under straight stalks, those microbes in the soil need some sort of nitrogen source to process the carbon. And if they don't have it there, then they will take it out of the organic matter in the soil. And you can actually deplete the organic matter a little bit. But if you plow under a nitrogen source, with your carbon source, you are not depleting the soil of any nitrogen." He no longer plants corn, due to fear of contamination by genetically modified organisms, but he still has complex cropping plans to "keep that balance in the soil."

Complex rotations are necessary in vegetable production when twenty-five or thirty crops are produced, not even counting numerous varieties (that is, "lettuce" counts as one crop, although many types may be grown: red leaf, green leaf, romaine, etc.). Phil describes one of his many California crop rotations: "We have a pretty diverse crop mix rotation. We are on a five-year rotation on alliums, so onions, garlic, shallots, leeks will adhere to a five-year rotation," and he notes the financial and ecological benefits of some crops: "I think now there is enough potential to make decent money on lettuce, so it is worth keeping it in our rotation. It is fairly quick. It turns your ground around fairly quickly. It has some benefits with weed control."

Antipesticides

Not surprisingly, organic farmers are vocal opponents to pesticides, both insecticides and herbicides. In Florida, Rob says that conventional growers "don't know anything about working with the resources you have instead of working with synthetics. All these conventional growers use everything: herbicides and chemical fertilizers . . . all the skull and crossbones products in the world. That is all they have ever used, so they don't know any different."

Joel cites some of his readings: "It's been proven in Third World countries. A little country in Asia, maybe Indonesia, was the seventh largest user for pesticides for rice. The government just banned it one year. And within two years, they had increased their production 10 percent above when they were using chemicals. And they have this farmer-to-farmer program where they teach one farmer organically and he goes and teaches other people. They have now reached a million farmers in their program. But the whole consensus was that once they were off chemicals they saved a huge amount on their inputs because they were not spending it."

Phil and Steve say that being organic may help them attract more and better seasonal workers. Steve says, "I think it helps a little bit. They know we don't spray. They know we are not like other farms." In Colorado, the Bensons all know of neighbors and family members who have been hurt while handling pesticides. Allen says, "I don't want to be messing with that stuff." Cliff says, "Farmers are brainwashed into thinking they have to have it." In Illinois, Joel and his family attribute his dad's Parkinson's disease, which claimed his life last year, to his work in pesticide applications. "It's from the chemical use and exposure to it."

Science

The scientific justification for organic techniques and the field comparisons between organic and conventional crops may be questioned in the academic world, but to organic farmers it is quite clear. In Illinois, Joel describes the problems with conventional corn crops: "You talk with all the plant genetics people and all these new genetics are being bred to uptake N, P, and K, but nothing else. There was a man who took comparisons. He took four varieties of open pollinated corn for feed analysis, and he took eight varieties of different Pioneer, De Kalb, blah, blah, blah – supposedly the good silage. And when he got into the feed quality, he did a gas chromatograph analysis for all of these. It tells the mineral content. The open pollinated corn, as far as feed value, I think it beat the conventional varieties by 600 percent as far as the nutrient quality." This affects livestock. "Think of the grain that you feed them. These conventional varieties – that is why you have to add all of these feed additives. Because the corn doesn't have it."

Regarding Florida citrus, Rob says, "Not that organic fruit won't freeze, but it will take a little bit more cold for a little longer. Conventional fruit has no mineralizing in it, and it will freeze up real quick. If you have a good organic program, it will withstand drought, and it will withstand extremes a little better than the conventional stuff." Mary says that people have tested their citrus with "a refractometer, and it runs sugar content and tells you different things about the fruit," and theirs is "consistently high quality."

Based on science and their own experience, farmers chose various techniques for organic production. In his California vegetable production, Phil says, "We have been putting in more insectory areas throughout the ranch. We have insectory hedgerows in our orchards. We don't disk, but we leave some vegetation and do alternate mowing. So all these are more habitat for beneficial insects. We put in some annual plantings for a beneficial habitat. We have alfalfa strips. So all these things in very small ways really add to the amount of predator control we have on pests." Phil has help with his scientific approach to pest management. "I have a person that walks the fields, in addition to me, two to three days a week. He has a Ph.D. in pathology. He has really developed his insect monitoring skills." Using their own skills, the scientific information they read, and scientific consultants, organic farmers work hard to keep current with methods of organic production, pest control, and fertility.

Environmentalists

Farmers, particularly those with a conventional farming background, tend to be leery of "environmentalists," whom they believe are uneducated about, and distrustful of, farming. Environmentalists tend to blame farmers for ecological degradation (often rightly so). Regardless of the ecologically balanced actions they undertake on their farms, even some organic farmers have conflicted views of an "environmentalist." Phil in California says, "My wife is an environmentalist, and she has had a big influence on me. I don't know that I would term myself as an environmentalist, although I am probably a lot more so than I was ten years ago. When somebody grows older, they think things out a little more. But I would say I am more an environmentalist than not an environmentalist."

Naioma in Colorado says, "Environmentalists don't understand. If you crack down too much on farmers, they will quit. Can we still feed the world?" Although she has strong reservations about the motives of environmentalists, she personally feels that "you cannot manage nature. You have to be part of it." These ecological concerns provide a glimpse into the complementing and contrasting relationships between organic farmers and the rest of American society.

SOCIAL FORCES

The success of organic farms is associated with values and ideals of society. Specifically, American culture has a great deal of influence on farming, food consumption, and rural life. Our government determines agricultural policies, organic certification rules, and research themes, all of which are shaped by American social values.

American Culture

Several aspects of our culture are very influential in organic farming. First, organic farmers disagree with how Americans support the conventional agricultural system, when in reality it is not sustainable. Second, public perceptions of organic farming itself are important. Third, Americans expect and demand cheap food, which further maintains the current conventional system, even in the face of ecological and rural social decline.

Conventional Ag

Conventional agriculture holds little promise. Organic farmers have recognized this fact and taken action to escape that system. As Phil describes the situation in California, it is difficult for a small-scale conventional farmer to survive. "I think conventional is tough now, especially being an individual staying in farming. You have to have a lot of luck, a lot of skill, and a financial backing behind you. I don't know if I would relish being in that situation. Conventionally, the marketing options are limited; you are relying on someone doing all of your sales for you. I think it is a hard field."

When asked about the future of conventional agriculture, Joel speaks about prices from his Illinois context: "It's tough to say. I really don't know. It's almost like they would have to be in financial ruin before they realize that some changes need to be made. It is completely different marketing organic than it is conventional. Let these people lose their money on options and all of that nonsense. Where else are you going to get an offer in a conventional system? Everyone is a penny or two away from each other. There is the price. Take it or leave it."

In Colorado, Cliff Benson feels that "conventional farmers are only concerned with increasing yield. How do you encourage a shift to organic? Lower production goals! Organic reduces yield per acre, but there are lower input costs, so you get a profit." So farmers first need to change this "high yield" mind-set.

Then, according to Steve, "as long as there are good marketing opportunities, organics will continue to grow. If margins tighten up, I think the transitions will slow down. That is what really helped it grow over the last few years. What are conventional corn or beans, even vegetables, worth? I feel there are still a lot of people looking for alternatives because conventional agriculture is just going nowhere."

Looking at the citrus market, conventional production is primarily for the juice processing, which allows farmers to cheat on quality compared with the fresh, whole fruit market. Rob says, "Florida is mainly geared for juice production, for that 'poison in a jug.' Once you have fresh juice, I don't know how you can buy that stuff; it is garbage. And it is all grown with chemicals, and then they cook the shit out of it. You'll have a hard time convincing me that there is any nutritional value in that. Plus half of it is green when they pick it. But the stuff tastes like battery acid. It's nasty." Rob continues to describe how large-scale conventional citrus production has a huge influence in the state of Florida: "Now there's a lawsuit about generic advertising. Basically all the advertising you see for Florida orange juice benefits about three or four corporations, and that is it. That ain't doing shit for the grower. It is all politics, and it really sucks."

With such strong feelings against conventional agriculture, it must be difficult for organic growers to be minorities within the current system. Indeed, it is interesting to ask organic farmers how they fit within their local conventional agricultural community – those conventional farmers wearing the agribusiness, chemical corporation–sponsored caps, sitting at the diner drinking coffee. Since organic farmers are so distinctly different and maintain obviously different field methods, they often feel like outsiders. But they are smiling as they look back in. In California, Phil says, "Am I part of that ag community? Probably not. I don't know if I would be even if I were conventional. I have some good friends that are conventional farmers, and we stop by the side of the road and talk about different things. I'm growing onions, and he's growing onions, so there are some commonalities. But certainly if someone is an organic farmer, I would have more to talk about."

Views of Organic Farming

In Colorado, where conventional farming rules, Naioma says, "Some neighbors think we are loony tunes for doing organic!" Her husband, Cliff, agrees. "Some neighbors think I don't have both oars in the water!" In Illinois, Joel says, "I never get the questions. Dad would always get the questions about me – 'What's he doing here? Why is he doing this? Oh, is there a market for that?' I had hairy vetch out in this field here, and people driving by didn't know what it was. 'So what is he growing? What is he going to use that for? Is there a market for that?'" Since his flame weeder is new and different, he figures the neighbors will ask about it. "I haven't heard anything about flame weeding yet. I don't know if they haven't mustered up the courage to ask, or what."

In upstate New York, Steve remains upbeat about his relationship to his conventional neighbors. He says he is "a little isolated, but not too much. Our neighbors are still neighbors. But it is nice to talk to other organic farmers about your concerns or marketing problems or whatever. You come back from the meeting charged up by people at the meeting thinking like you. Something different from the usual crowd I see here around home. There was an auction here the other week, and there was one organic grower there. It was good to visit with him, but that was it. No one really wanted to talk to us from the conventional side."

While farmers in other geographic regions are distinct, different, and perhaps the brunt of a few jokes by conventional neighbors, in California things have moved ahead. Based on his experience in California, Phil sees that organic farming and organic food are becoming more generally accepted. "Maybe not the stores in small towns, maybe not stores in the Midwest, but certainly stores in this area. Even conventional stores are carrying a little small section of organic. And there is some expansion going on all the time. With more areas in production, with better supply, with more year-round supply, chains and stores can afford to get back into organic, because they know there will be some consistency to it. There is a synergy going. As more people get in and there is more supply, there might be some depressed prices as far as a grower is concerned. As the prices are lower or closer to conventional, then a lot more people have the ability to buy it. And then chain stores will carry it, and more people can buy it." Such mainstreaming of consumption will likely filter slowly into the rural countryside. In the meantime, these independent innovative organic farmers just keep their chins up and pride intact.

Cheap Food

It is unfortunate that Americans truly believe they have a right to cheap food. Phil describes conventional food prices, based on his California experience: "I think sometimes our food, especially in the vegetables, gets so cheap. There is just so much supply that comes on at a certain time that people fight each other for the sale. Then the prices just plummet, and nobody wins with those prices. The consumer may win in the short term. Whether they buy lettuce for 39 or 59 or 79 cents, what the farmer is getting for that is so small."

Prices are low to appease consumers, who are apparently satisfied with "pretty" food, even if it is of poor quality and taste. In Florida, Rob says, "You go to the supermarket, they want stuff that looks good. They don't really care so much how it tastes; they just want it to look good. And Wal-Mart, they have the ugliest-looking shit I have ever laid my eyes on. It is probably nasty eating, too. Have you ever seen the fruit they have right here in town? It is pitiful. I can't for the life of me understand. If I were a buyer for a major corporation and I saw shit like that going on my shelves, I would be freaking out. I would be backcharging, writing credits, and I would tell them, 'I'm not giving you a nickel for that scrap.'" But Mary understands the consumers' viewpoint. "I am the one who buys the groceries. I'm the one who looks at the apples. If there is an ugly apple there, I don't care if it is certified or not, I'm not buying it. As a consumer I am going to buy the pretty stuff." Of course, she won't sacrifice taste, either, so there are multiple concerns in terms of food quality.

Summarizing all these consumer demands, Phil notes the variation between conventional and organic food: "I hope consumers realize how important their food supply is. At least organic consumers perceive they are getting more, and they are willing to spend more. I think they realize the extra value. Either they think there is value because there are fewer pesticides, or they think that organic farmers are doing something better for the ground long term. So people will pay more." The question is whether organic food will follow in the path of the entire U.S. food system and become so cheap that quality doesn't matter. The organic marketing specialist in California, Terence, sums it up: "Americans always want everything cheap. The goal would be: how can I put organic in people's hands as cheaply as possible? Rather than: how can I treat the soil and the earth in the best way possible – so this land will be good for 10,000 years?"

Rural Regions

Speaking from experience in grain production on the plains, Naioma points both at farmers, for their inability to adapt, and at the agricultural system. She says, "Farmers have a 'prove it to me' mentality. I've read that it takes farmers twelve years to adopt a proven practice! It's true. It's also driven by economics; bankers control what we do. In the late 1970s, credit was easily available. Bankers went door to door offering loans. Then crash! The 1980s bloodbath in agriculture. In 1983 to 1985, foreclosures were high. We lost lots of people from agriculture. Then it evened out for a while, but now it's bad again. A lot of farms are going under."

In Corn Belt agriculture, Joel notes, "Everyone around me, I think, is farming over a thousand acres. So I have set as one of my goals to make as much money – net – as the guy farming a thousand acres. I hope I can attain it. I know I will. That is my goal on three hundred. And I know that is going to be possible. But that is the beauty of it. I don't have to buy bigger equipment to farm more land." He won't try to keep up with the "bigger is better" mentality of conventional agriculture, which depopulates rural regions. Instead, he's making it on less land.

As neighboring farms are much larger than his, Steve speaks of the land pressures in upstate New York agriculture: "Since we are in a fairly competitive area for ground, vegetables are a way to expand without expanding our land base because they are higher in value. Land is very competitive, especially good vegetable growing ground. Both the fresh market people and the processing people, I have to go out and compete with for ground."

In Florida, Rob describes how the rural regions have changed: "Like the Florida Citrus Commission, the rules are rigged. And everything is geared to help out the Big Shots. They ain't looking after the little people. That is long gone. Little packinghouses like this, there used to be thirteen of them in Crescent City. I am the last one left in all of the county. It used to be small growers were with twenty to forty acres. Now small growers are a hundred to two hundred acres. They have done a good job of pushing out the little guys."

In many regions, land prices are skyrocketing because of the encroachment from suburban sprawl. Joel knows this issue, as the suburbs of Chicago are pressing further out in his direction. "People by the city are getting bought out for these huge amounts – $20,000 and $30,000 an acre. And they just move west and find a nice place. Farmers sell out and then have money to buy farmland further out." Joel is renting his farmland from his uncle, and he is worried. "Unfortunately, it doesn't make me feel very good. People come into his office and say, 'Is your farm for sale?' He tells them no. So I don't know. We will farm until we can't farm here anymore. It would just be nice if my son wanted to farm, that he would have that chance." But it's an uncertain future for many rural areas, as land prices increase and farming opportunities decrease.

Policies and Information

Organic farming is influenced both directly and indirectly by several components of U.S. agricultural policy. First, the new USDA National Organic Standards may impact markets and production. Second, organic certification influences farmers' techniques and ability to market through distinct channels. Third, most national and state agricultural policies are not tailored for organic farmers, but they often still affect them. Finally, policies and programmatic funding greatly influence the types of research conducted and whether the results are relevant to organic producers.

USDA Standards

The USDA National Organic Standards went into effect on October 21, 2002. Organic farmers have mixed feelings about the potential of this new policy, the motivations behind it, and the consequences it will have for them. In general, the USDA standards are seen as benefiting larger organic producers the most.

As Terence notes, "The national standards are set up for large companies that are exporting. That is what they are about. Small farmers tend to sell locally. They tend to know a lot of the people they sell to at the farmers' markets or the distributors. There are direct relationships and trust. And the problem prior to the implementation of the national law was that somebody from another country would not know the standards to which the growers were growing." Rob, as always, is blunt: "I'll tell you, I have a real problem with these national standards. They are really out of touch with reality. And everything is written by and for the big people." Joel also has harsh words: "That is typical. Let the government get in there so they can screw it up." Mary says they are considering looking for a stricter standard. "It seems like organic is getting away from organic, so I'm going to have to go a cut above to distinguish myself. It is credibility. The things they are letting organics do should not be done. But that is what happens with national standards. So we're going to have to find something a grade higher."

Many organic farmers agree that the additional paperwork is the worst part of the new standards. Joel says, "Our farm questionnaire went from three pages to seventeen pages because of the national standard. That is for everybody. So it has added a huge burden of paperwork. What bothers me is that the USDA doesn't set up any kind of regulatory thing. Nobody is checking anything. So it is 'buyer beware.'" But people within the organic industry may force the establishment of better controls. "I heard the Organic Trade Association was going to sue the USDA because it is a law. They don't have anybody going to the store and getting a chicken and taking it to the lab to sample it: the enforcement."

But Steve says that many organic farmers are complaining too much. "There has been a lot of hoot from the small farmers who maybe didn't get certified before or got certified by groups who didn't have much of a program. They are complaining about all the new paperwork and the fees, but our fees haven't really changed with the new rule or the paperwork we need to do. This year we saw no real difference." But his certifying agency has been phasing in the new standards for several years. "The NOFA–New York program was ready to go with the new standards. They knew it was coming on, and our standards were tough enough that we just didn't really see a change." Even Steve admits that he probably won't have any benefits from the national standards. "I don't think so. The only thing that might help us is maybe with the grains. As long as we are certified by somebody accredited by the USDA, they have to accept it. I think we are getting a little benefit from that."

Certification

Related to the new national standards, certification of organic farms is a key for most farmers' marketing outlets. The paperwork and rules take a great deal of time and can seem particularly aggravating, since farmers need to be outside working in their fields. The national standards may have created more paperwork, but basic certification was already paper-intensive, and some rules seem inappropriate. Allen, in Colorado grain production, says, "I'm going to stay organic for the time being, but I don't know. The paperwork. They are getting more and more strict every year, adding five pages to the stack of rules each year. Half the people making up the rules haven't ever stepped in a wheat field in their life." Red tape aside, organic certification is important to most farmers. That certification allows the Bensons to sell their wheat for a higher price than noncertified grain.

Another concern with organic certification is whether the rules meet a high enough standard. According to Mary, "They have a lot of exemptions. There is a whole list of products that you can get an exemption for, that, as far as I'm concerned, once you use them you are no longer organic. We've been calling for years, trying to raise the Florida standards. What is really sad is that they are letting organics gas [using synthetic ethylene]. Citrus growers don't need to gas. An orange will turn orange all by itself on the tree. It naturally turns, like a banana will turn yellow. We've petitioned Florida Organic Growers to disallow gassing of citrus. And since some people swing a lot of weight in the business, we were turned down. They accepted gassing. They said they could not supercede national standards. We've been in contact with other certifiers, and they said they would supercede the standards, so we're thinking about getting certified by someone who wants to keep the integrity of the business."

The problem with minimal adherence to the organic certification standards is real, but Steve has a logical view, as always. "That is a concern. We are hoping the standards we are following and the inspections are just as tough for everybody else. We are worried with things like biotech and pollen drift, very concerned about that. They are probably more concerned where all of the corn and soybeans are growing. But we are just hoping people will continue to pay some kind of premium for organics because if it isn't profitable it stops being fun real quick."

As more and more organic food is imported (or grown abroad for U.S. corporate farms), there are additional concerns. As Terence, the California organic marketing specialist, explains, "In Mexico and other countries that don't have an organic sticker community, they would be the most likely to have fraud occurring because there is no one to be watching locally. The common Mexican citizen couldn't care less whether the produce going north of the border is really organic or not. Why should they? They are looking at us as having way too much money and using way too many resources. They just don't have incentives as far as I can tell, whereas in the United States the local people are watching. There have been incidents where organic growers have been turned in because somebody thought they were spraying a prohibited material or that they were cultivating land that was not properly certified." On the other hand, in the United States, "I think you are more likely to find minimal adherence to standards. The organic movement originally started as something ideological where people were always trying to improve and drive to a higher level. I think as it gets more commercialized you will see the opposite happen. People are saying, 'What can I get away with in terms of minimal adherence? What is the least amount of compost I can put on? Or the least amount of cover crop I need to grow?' "

Government

In addition to specific organic policies, agricultural policies in general affect organic farmers. Whether they participate in federal agricultural subsidy programs or not, these organic farmers voice clear opinions about federal agricultural policies. Cliff explains that "tax dollars go to agriculture, so food prices stay low. If you stop subsidies, then you should provide a minimum price per bushel so farmers can make a living. Everyone else in America has a minimum wage. Farmers need a minimum wage or minimum price per bushel. Farmers need that security."

Joel calls it the "Farmer Welfare Program." But he admits that he participates with some crops to please his landlord: "Yes, but we won't change our practices for the farm program. If the farm program will fit into what we are doing, then fine. I am not going to put in all of these corn acres just to get my money. We are still able to qualify the small grains, so that is what we are doing." But Joel realizes, "I am a hypocrite. I bad mouth it and then stick my hand out for the check. None of my decisions are going to be based on what farm programs there are. I look at it as a bonus, I guess. So, if it is there, fine." Joel continues to point out the irony: "I have done all of this environmental stuff on my own without any government help. Like putting buffer strips next to my creek. I put in an alfalfa and grass mix so I can hay it a couple of times. I have done a contour and I have done strip contouring and then I have done a Christmas tree creek bank stabilization project with no government help. But, you know, with the cover crops and all that keep the ground covered, we don't get paid for some of the good things that we do that could really help." Conventional farmers typically get technical advice and cost-share payments to implement these ecological restoration activities that are often commonplace on organic farms (at no cost to taxpayers).

Joel states what many other farmers feel, that the subsidies are due to collaboration between the government and agribusiness, working to keep commodity prices low. He says the future is fairly certain: "Government programs? They won't stop it. I don't think anyone in Washington has the guts to stop it. They don't have the guts to overhaul the other welfare system, let alone the farmers' welfare system. It's a shame. But these big companies are not going to let them end. That is why the farm program is teaching us not to think for ourselves. So what if you get a poor yield? You get another $10,000 from the government."

In Florida, Mary provides an example of how government policy has trailed far behind the needs of organic producers. "Back in the sixties and seventies, the government would not recognize organics. The Citrus Commission said it could not be done. In the early eighties, they tried to start an organic citrus growers' co-op here in Florida. The Department of Citrus would not recognize organics, so all the stuff sent out of here on the train was bootlegged out. You were not allowed to send your stuff across the state line because you weren't spraying it or fumigating it. So all of this stuff went out of here on backroads to the Northeast Co-op." Now certification agencies are following national standards, and we hope that the acceptance of and support for organic production has been established across the United States.

In addition to dealing with organic inspectors and certification, some farmers have other inspectors as well. For example, Mary and Rob must have their packinghouse inspected by the Florida Department of Agriculture, and they must have their citrus grade inspected by the USDA before it can leave the state. Mary jokes about the USDA inspectors: "All them love coming up here to my house because they never know what to expect. Last year they ran these tests. They juiced up these oranges and put a whole bunch of chemicals in there to see what the sugar level was. This one guy juiced all these oranges, had this big bowl of orange juice, and he set it on the steps of my packinghouse. Went to his car and by the time he got back, the dog had drank his orange juice! He was so mad. My dogs love oranges! Both of them do. They beg for them. And I thought, 'I won't sell this load in the state of Florida, and I won't get the paperwork,' but it was just too funny."

Rob is more negative about the value of these inspectors. "Who knows? Like our Eagles' Nest Brand is U.S. number one, and they want to make sure that is what is in the box. They have to come out and inspect the fruit and put a USDA stamp on it. Unfortunately, I have to pay the USDA. It seems like everyone has their hands in your pocket."

In Illinois, Joel has additional government inspectors: "All of our meat is frozen. Yes, we've got five freezers. Having our broker's license, a man comes to check it once a year. Although I am wondering if he retired because the state of Illinois offered early retirement. I haven't seen him for a while. You fill out the forms, write him a check, and he checks your freezers. Are they cold? Yes. Are they processed where they are supposed to be? Yes. Check who you are processed from and make sure it is from an inspected plant, which it is, so there is no problem." Paperwork and inspections are routine for organic farmers.

Research

Organic farming is management intensive, and farmers need information for their diverse operations. But when Joel is asked if his college degree in agriculture helped him learn about organic farming, he is adamant: "No, not at all. You won't find anything from universities about organics. They're so far off base. As far as universities getting in touch with really sustainable practices, you can forget about that." Naioma thinks that universities should "switch research to sustainable or organic agriculture. At all land grant colleges, research is funded by chemical companies. Information from universities for organic producers is null and void." Joel agrees. He says that if we had a wholesale shift to organic methods, "all the universities would have to find work, because they are all, for the most part, chemically funded." Phil notes that even in California, "information can often be difficult. Up until recently there hasn't been much interest in researching organic, especially in organic systems." From a New York state perspective, Steve says that information from "extension and university people" has improved over the last ten years or so.

Based on his Florida experience, Rob says, "It is a shame with modern agriculture. They teach it in the schools, and the kids that are coming up in it don't know anything but chemicals. It is all politics. All this agricultural news I get, like *Citrus Grower*, *Florida Citrus Industry*, *Grower and Vegetable*, it is unbelievable what is in there. Everything is written by and for the chemical corporations. You will read about field trials, and then at the bottom it will say this product was used and that product was used. You ain't never gonna read about a bad field trial. Whenever there is a writeup about a certain product, the article makes it sound like the next best thing that ever happened."

Joel speaks with personal conviction about university research on organic farming, since he has two degrees in agriculture. "That's all they want to see: more university research. The guy that has been doing it on his own farm for twenty years is not reputable. I took a nematology class. The professor said, 'You have to spray this and this and this.' I said, 'No, you don't. Use compost and crop rotation. You don't need any chemicals to kill nematodes.'" Joel described how farmers have implemented organic techniques that work. They even take soil samples or count pest species before and after. But these studies are not accepted by agricultural colleges, "because it is not replicated. The farmer doesn't care that it was replicated. All he cares is that he had a problem and you solved it, relatively inexpensively and using no chemicals." The overall success of organic farms hinges on more than just information and the values of society. It is also associated with individual farmers' independence and innovation.

PERSONAL CHARACTERISTICS

Personal motivations and characteristics drive farmers into action. Organic farmers are independent and innovative as they pursue the path of organic farming. Thus they are reaching forward, seeking a better future in agriculture. At the same time, their family and their roots, so to speak, often remain firmly within the farming tradition.

Independence

These farmers are independent in their search for information on organic methods. Part of this independence has meant that they must be willing to assume risk rather than remain content within the failing system of conventional agriculture. Although they take risks, they are not foolish with their money, preferring instead to maintain very low debts. Their strong independence means that they do a lot for themselves, so they are extremely busy people, with multiple simultaneous farming tasks.

Information

Information on organic methods is not readily available from the usual sources of agricultural information. In Illinois, Joel states what most organic farmers say: "You basically have to go to other organic farmers. I go to all of the meetings. That is where you get all of your information about how to do things. There is this one man who I call my organic coach. I go to the meetings early and find him, sit down, and tell him what I am doing." In New York, Steve agrees. "My main sources of information seem to be other farmers. I've found since we started going organic in 1990 that there seems to be more extension people out there with interest and knowledge on organics. Local USDA people have no knowledge on organics and show no interest in learning as we're the only organic farm in the county. We're always looking for info on controlling a couple of insects."

Sharing information can have positive and negative aspects, however, as Mary describes helping a new organic farmer in citrus. "People take up years of our time. There was one that took up three years. At least twice a week, four hours a day. We don't mind helping you. But listen, this is getting a little old. And then as soon as they get certified, guess what they do? Stab you right in the back, and I don't have time for that."

In Colorado, Naioma describes a common issue in terms of sharing information on organic markets. "It's not cut-throat. It's just we work hard to develop markets, so we don't go there. We don't talk about it." Their phone bills are high because they contact so many various brokers, distributors, and wholesalers to reach a price and product agreement. She says, "In this atmosphere you guard your buyers. Even my brother and I don't share this information – so what are you moving, millet or wheat? There is no sense in me asking, because I'm not going to tell him next time he asks me what I'm moving. Yes, it is very, very competitive."

In California, Phil agrees that information is both shared and kept personal. "I think organic farmers used to share a lot more information. It is becoming more like conventional, where people are guarding their information. There is competition out there. People are being careful about being free with their information. In the early nineties, there was a lot more camaraderie; there was a lot more information sharing. There are still things that come up. It would still be nice to share information, but at least we have a lot of stuff we have learned over the years. When problems come up, we can usually figure some things out on our own."

Risk

Clearly it takes a certain acceptance of risk to try something as different as organic farming when all your neighbors are conventional farmers. Especially with information on organic methods relatively hard to come by, these farmers must be independent and confident. The Bensons agree that risk is part of the game in organic farming. Cliff says, "Some of the decisions I made when I was younger and, as I look back at it, really worked well. It would scare me to death today. But I just jumped out and did it because that's the way I was gonna do it. And it worked out good for us."

Phil describes the link between risk and diversification in terms of his California ranch: "We have minimized the risk by having such a diverse operation: having diverse marketing, being in charge of some of our sales, being able to get some of the extra dollars by doing some delivering." In Illinois, Joel still gives advice to other organic growers, and he notes that there is a link between information and risk. When another farmer comes to him for advice, "I just tell him what I am doing, what works, and what I have read that other people do, because I have done a ton of reading. But even after all that reading, every farm is different. So what works for me might not work for him. You basically have to go try it yourself. I have put everything on the line. Either we are going to make it fly, or we'll get out of farming. So that is my attitude. It's all or nothing."

Low Debt

Successful organic farmers tend to be quite conservative with their money. Phil explains, "We do a lot of budgeting and forecasting. Even as well as that goes, there have been times in the last three or four years where things have been out of our control. We have had times that we had to borrow money. We paid it all back, but we've gotten up pretty high on our line of credit. That is just more to worry about. When you hear someone going out of business farming, that kind of feeds on you a little bit." Rob says that he has grand plans for his Florida grove, but "I'm broke. That is one of my problems. I have never had any money to dig in on. I don't like being overextended. When you have debt load up to your neck, that makes your pressure even heavier. I am pretty much happy with what I got. I just wish I had a little more cash." In New York, Steve says simply, "We are frugal, or we try to be."

Just as it is important to keep debt low, organic farmers are conservative in their equipment and machinery purchases as well. They are not motivated to buy the latest combine or the biggest tractor. They buy what they absolutely need, often used machinery, and they are likely to do all the mechanical work themselves. On his California ranch, Phil describes: "We try to buy a tractor that has multiple uses for it. We have generally bought used tractors – basically out of rental returns. That is how we bought most of our tractors. Most of them we would put a lot down on them and then pay them off."

Joel says, "It is my goal to save up enough money to at least not have to borrow money for operating expenses. And that will happen. It won't take too long for that to happen. Then we will just be borrowing money for the cattle. I am buying out my dad's equipment on a lease agreement, so I will pay him so much a year to use it, and every year, a little bit more gets turned over to me. There is no way you can just come into it and say, 'Oh, I think I will farm' and buy all your equipment. And I haven't bought any new equipment. I don't need any of that. I do my own repairs, everything – rebuilding motors, transmissions, etc." Likewise, Allen does all mechanical repairs on his Colorado farm machinery and even builds his own specialized equipment, like the grain cleaner and container loader that "blows the grain up into the containers" for rail shipment.

Busy

Seeking information, growing and marketing diverse crops, and maintaining their equipment, these independent farmers are very busy. They are experienced at multitasking. In fact, Rob noted that the only time he is indoors is after sunset. "And I don't know if it is going to slow down. All I have on my mind is taking care of grove maintenance." So even after harvesting and shipping season is over, he's busy with his groves. Mary says, "A lot of times, I load 24/7. Because if I've got product here and it needs to go to Boston and a guy wants to pick it up at 3:00, it isn't his fault that there was a wreck in Orlando and he couldn't get here on time. Or that he got to some packinghouse and the product wasn't ready. I'm loading 24/7. I'll load in my pajamas. I want that product on that truck and out of here."

Steve says some seasons are busier than others. "In the spring we are pretty hectic. But in mid- to late July, things start slowing down before the big fall push." Joel is particularly busy in the fall when the farmers' market is in session, but he remains active all year with meat and poultry sales to restaurants and buying clubs and the multiple grain sales to various buyers. For the Bensons, their grain production is busiest in the summer when preparing the ground, planting, getting equipment ready, and harvesting must all be accomplished in a fine-tuned sequence. Plus all of these crops must be sold, and most often there are multiple buyers and brokers for each. Phil's California vegetable farm is in operation year-round, although it is a bit slower in the winter. Even then they sell shallots, onions, cabbage, squash, "and crops that can be pretty successful in winter harvest."

In addition, organic farming methods are more labor-intensive than conventional agriculture, as humans often take the place of chemicals. Rather than spray a pesticide, organic farmers carefully monitor the occurrence of pests, and then may decide to rotary hoe, cultivate, pick bugs, flame weed, or even use a bug vacuum. Each of these techniques requires time and planning. The paperwork and inspection process for organic certification add another time-consuming component to these farms.

Innovation

Organic farmers describe how their success is partially due to their willingness to try new things and experiment on their farms. They are also motivated by an open mind and the desire to learn.

Experiment

Since information on organic methods is not readily available, and because many problems and solutions are locally based, many organic farmers conduct their own agronomic experiments. This on-farm experimentation runs from small changes in timing of planting to full-fledged side-by-side trials of a crop grown with different techniques, to innovations in the types of crops they decide to grow. The Bensons, in eastern Colorado, grow the only dryland alfalfa in this region. Cliff explains, "The neighbors think it's crazy to grow alfalfa here, since it receives only twelve to fourteen inches of precipitation per year." But according to Naioma, "The soil needed help when we bought that land, and alfalfa helped and grew pretty well." Naioma described one of her husband's innovations. "He is trying new flaps on the back of the planter to help set the seed better and have less [soil] compaction." It is different from typical farm equipment, so "the neighbors just laugh and shake their heads."

Phil agrees that a lot of information comes from within your own farm. "There has been a lot of experimentation, a lot of scratching our heads and figuring things out for ourselves. We keep our hand in it, continue to experiment, see what we come up with." Experiments are common on Phil's farm: "Cabbage is one of the crops that we have an insect pest problem that can be pretty significant without a lot of good control. In the winter the problem is your beneficials are not as active, and the cabbage aphid can get you in the winter. We still need to do research on it."

In Illinois, Joel is conducting field and livestock feeding experiments. He describes his latest project: "I am dabbling in verma composting with worms. I am working with some people out of Wisconsin, in Milwaukee. We are going to be doing replicated trials using verma compost and preplant fertilizers and verma compost sprays." Spraying fertilizers made from dead worms? Yes, because there are always new ideas sprouting up on these organic farms.

Learning

In terms of educational background, organic farmers vary. Some have studied agriculture in college. Conventional agriculture is the only type taught, according to Phil, Joel, and Steve. Several have mechanical or construction backgrounds, such as Allen, Joel, and Rob, which is helpful in maintaining farm equipment. Rob says, "Hell, I'll grow anything I want to grow if I put my mind to it. Citrus is what I know the best." Mary says, "I'm not book smart. I've got common sense. Everything I have learned is self-taught. So I am not as likely to forget what I have learned."

Steve is from New York, but he has "the equivalent of three years of ag education at Iowa State. Most of the courses I took there were centered on livestock and ag engineering." Joel says that after high school he went to a community college "and took an ag mechanics program, and then I went into the Peace Corps in St. Lucia in the Caribbean for three years. Then I came back and got my engineering degree and then came back to the farm." Phil got a degree in ag science and management in the seventies, but "didn't learn of organic until probably the latter half of the eighties."

Regardless of their formal education, these farmers are learning constantly, as they fine-tune their farming techniques and tailor them for their specific location. They actively seek information to answer their specific cropping and livestock questions, and they apply new information to their own on-farm experiments.

Tradition

Allen Benson is the fourth generation to farm this part of the northeastern Colorado plains. When asked, "Why are you in farming?" his dad, Cliff, explained, "I like it better than anything else I've done." His mom, Naioma, replied, "Tradition. It's our land, our life. For the love of the land. It is twenty-four hours a day. Farming determines your life: when you eat, the number of children you have, when and if you can take a vacation, and so on. It's an occupation and vocation."

In Illinois, Joel grew up on a farm, and he wonders how long his family will stay in farming. "My son is twelve and my daughter is eight. I don't know. He likes tractors and stuff like that, but he doesn't say much about it yet. They are young, and I won't push them into it. The only problem is when you don't own the farm. That is always a question in the back of your mind: are they going to get a chance on this place?" The region is starting to feel the pressure of suburbanization creeping westward from Chicago, making land prices soar.

Steve's family is firmly rooted in upstate New York, although their farm has changed over time. "I see myself as a farmer first and an organic grower second. We grew up with the livestock farm." His parents bought this land fifty years ago; the family farming tradition continues here.

Evolving

A key component of these farms is that they are not stagnant; rather, they are constantly evolving. New crops are grown, different markets are tapped, and particular field methods are employed. These farms are moving forward rather than being stuck in the past. Each farmer has plans for the future.

About their Florida citrus, Rob says he'd like to have more time and money. "That would be nice. Then I can plant my back field heavy in grapefruits and tangerines. Maybe a row of this and a row of that. I might plant a row of peaches and one of plums. I had a few peach trees here that, oh, I had never had a better peach in my life." Mary knows that marketing their citrus is an ongoing process. "Never become complacent with what you have. I've been in the business long enough to realize that next year three of these people may not even be in business any more. You can't depend on this." In Illinois, Joel explains how he is trying to direct market more now: "I am trying to be 100 percent direct because it shields you from these stupid commodity prices. We are not there yet, unfortunately, but we are working on it." In California, Phil hired Terence, who is a marketing specialist, and they have moved toward "setting up a local delivery system" and away from wholesale markets. Allen, in Colorado, is keeping more profits on the farm, as he developed new equipment to clean and load his own grain. "That has made a difference with our sales." And Steve's farm has evolved, shifting from livestock to vegetable production in 1990 and beginning the CSA in 1996. They are planning to increase the CSA or, as he says, "grow the CSA," since it has become an important part of their farm's success.

A FUTURE OF GOOD GROWING

Innovative family organic farmers, seeking to confront our failing conventional agricultural system, conduct their own agroecosystems research and discover how to grow diverse, high-quality crops that can be marketed directly for a better profit margin. These successful farmers have discovered how to overcome many barriers in order to make the most of all opportunities. Multiple economic, ecological, social, and personal factors are clearly influential in the success of organic farms across the United States. The next question is how to move forward and create a rural landscape that promotes the very characteristics that support organic farms.

6

Organic Farming in Our Future Landscape

Agriculture constitutes the last vestige of small-scale enterprise and widespread ownership of productive assets in our society. It offers our best and most important opportunity for environmental improvement, because we know how to produce food in far less destructive ways than we now do. And at the bottom of the farm crisis, with broken pieces of the farm economy laying at our collective feet, we have the chance to pick things up and rebuild the way we want – stronger, truer, and fairer. In agriculture we now have a good chance to do things right. We have choices.

– Marty Strange, *Family Farming: A New Economic Vision* (1988)

rganic farming provides ecological and social benefits, and innovative organic farmers can be successful in agriculture today. "The fact that organic agriculture has become competitive with conventional agriculture in so many different situations, in spite of its relatively depauperate research and extension infrastructure, is a testament to its potential" (Lotter 2003, 104). Looking to the future, we must support family organic farms that are good for both the earth and its people. At the same time we must ensure that the popularity of organic products does not relegate organic agriculture into the industrial model of large-scale production and corporate-controlled distribution that hurts farm families. The best way to encourage a transition to a family organic agricultural system is to (1) set clear and reasonable goals for organic farming, (2) take a stand as an advocate for organic methods, (3) conduct agroecological research that is relevant to organic farmers, (4) establish appropriate policies at the national and state levels that specifically target organic farmers, and (5) work to protect organic farming from corporate interests. Luckily there are many people talking about these same issues, so the following elaboration of these five topics draws from diverse sources to help us chart our future course of action.

GOALS

We, as a society, must decide what we want and expect from organic agriculture. There are many possible goals, both ecological and social, as detailed in chapters 2 and 3. Organic agriculture could improve soil resources and provide quality topsoil for future generations. By reducing the use of agrichemicals, organic farmers could protect our surface and groundwater supplies. Or organic farmers could plant special heirloom seeds and develop crops to diversify the genetic pool of agricultural vegetation. Biodiversity could be the focus of organic farms, and these farms could be the basis for protected natural habitats in rural areas. Organic farms could improve landscape quality by providing visual and sensory variation in rural areas. Organic farms can act to preserve rural areas with picturesque farmhouses dotting the aesthetically pleasing countryside. Or organic farming could be charged with the mission of changing the industrial food system by building alternative marketing channels. Organic farmers could also establish local food systems that aid in rural development and help us "save the family farm." Organic farms may help attract attention to rural areas and educate our urban society about food production. Agrotourism could bring city-folks out to the countryside to learn about organic farming. And these are just a sample of the many things that organic farms could achieve.

But can organic agriculture really do so much? Which of these options is most urgent and most widely accepted or attainable? Our society must answer the question: *what do we want from organic farms?*

First, we must acknowledge that organic farms cannot be expected to do more, in terms of ecological integrity and social change, than other farms operating in America today, unless we provide substantial monetary support for them to accomplish such high goals. We cannot expect organic farmers to step in and rescue our rural natural resources, save the family farm, and improve social relationships within agriculture. They are and must be, first and foremost, a farming operation – that is, most of their time and effort goes into producing crops and selling them, not rescuing society. Second, organic farms should not exist solely to be singled out as examples of special farms within the oppressive framework of conventional agriculture. Rather, we can work to build a future in which organic farms become the norm. Right now we can encourage local organic farms by buying at regional farmers' markets or CSAs, and we should encourage the acceptance of organic farming methods by buying organic products whenever possible. Finally, we must carefully educate people about organic farming methods and the potential of organic farming. *The Organic Foods Sourcebook* (Lipson

2001) provides a concise overview of the key topics in organic food, information sources, and markets. This is a good first step for anyone wanting to learn about the current issues in organic farming. *The Real Dirt* (Smith and Henderson 1998) presents practical information on soil, pest, and crop management on smaller organic farms in the northeastern United States, and includes photos and entertaining comments from farmers.

Further reading on related topics includes *Fatal Harvest: The Tragedy of Industrial Agriculture*, which illustrates both problems and solutions (Kimbrell 2002). The vast ethical, technological, and ecological impacts of industrial agriculture prove that it cannot safely feed the world, thus the book also presents "revisioning agriculture" that outlines ways to confront the "power structures behind the industrial agriculture system." Possible solutions include shifting toward organic production that is outside corporate control – "organic and beyond," as the authors say. A key component of change is keeping the integrity in organic production, maintaining the ecological stewardship of organic methods, and creating an alternative food system based on organic farming.

For all of its benefits, we must acknowledge that organic farming is not perfect. One problem is that organic agriculture is following in the footsteps of large-scale conventional agriculture, with a handful of companies controlling national distribution. But this can be thwarted by consumers committed to buying from local family organic farmers. Some dishonest farmers may try organic methods just for the short-term profit they think they can earn, but they won't last long, because successful organic farming requires commitment. Trying to do organic farming halfway simply won't work, since certification will weed out violators. Sure, there are problems, but what is our other option – the end of organic farming? Is that what we want? No, because at least organic farming provides the ecological benefits of reduced pesticide applications. But we must fight to ensure its integrity.

A benefit of organic farming is that, as a certified system, consumers know what the criteria are. There are no fuzzy or overused concepts of "sustainable" or "integrated" production here; rather, "certified organic" farms are free of prohibited synthetic agrichemicals and do not use genetically modified organisms. But this should be the starting point, not the terminal goal, for organics. As Hightower states in *Thieves in High Places* (2003, 200–201), "The USDA label is a step forward for the environment and our health. But the label is only the first step." He notes that organics must deal with deeper social and ecological issues so that organic farming is "centered on the culture of agriculture." He warns against "the corporate grab" in organic products that is "nothing more than profiteering dressed up in a green suit."

Organic farming should allow family farmers to earn a fair price for their crops without the government subsidies and bailouts that prop up the current agricultural system. Finally, organic farmers as a whole are innovative and outgoing and tend to be willing to educate others, which makes them ideal for teaching the value of healthy soil, organic food, and rural life to the rest of us. It is clear that we need organic farming and that it may provide us with a future agricultural landscape of ecological integrity and social stability. So let's aim high but set realistic goals and strive to reach them in our generation.

ADVOCACY

It doesn't take much digging to find information on the hazards of conventional agriculture. Look at the Environmental Working Group (EWG) Web site at http://www.ewg.org/foodnews for the "Produce Scanner" to figure out what pesticides you are eating daily. This "Point and Click Produce Aisle" tallies government data on pesticide residues common to U.S. fruits and vegetables (Environmental Working Group 2004). After you are overwhelmed by the numerous unpronounceable chemicals that you're ingesting, you can read the EWG's recommendations for the "Foods You'll Want to Buy Organic." As of 2004, for example, the types of produce listed are apples, bell peppers, celery, cherries, imported grapes, nectarines, peaches, pears, potatoes, red raspberries, spinach, and strawberries. These common foods have high levels of pesticide residues, which pose health concerns for consumers – and this is not even considering the ecological concerns of pesticide use or the social destruction of rural life due to the conventional agricultural system, as discussed in chapter 1.

Several recent books provide clear examples of the destruction wrought by industrial agriculture. *The Forgotten Pollinators* by Buchmann and Nabhan (1996) is both a natural history and cultural history that describes recent human-induced declines in pollinators (bees, bats, butterflies, and birds, to name a few); one-third of our food comes from plants that rely on these pollinators. The authors note that billions of dollars in agricultural crops would be lost if declines in pollinator species continue. The reasons for this disruption of pollinators are habitat destruction and deaths from pesticides – two factors resulting partially or fully from industrial agriculture. In the long term, "the fields and orchards that sustain our food supply should never become too far removed from wildlands, or their yields will suffer" (223). Natural landscapes help maintain the stability of our global food supply.

A comprehensive view of ecological and social risks associated with chemicals in our environment, with some attention to Corn Belt agriculture, is found in *Living Downstream* (Steingraber 1997). In a reader-friendly style, Steingraber unites ecological and health data to piece together a picture of landscape and culture that reveals alarming news about cancer. She says that researchers focus on genetic and lifestyle causes of cancer while ignoring environmental exposures, but "lifestyle and the environment are not independent categories that can be untwisted from each other" (266). "After all, except for the original blueprint of our chromosomes, all the material that is us – from bone to blood to breast tissue – has come to us from the environment" (267). And indeed pollution of our air, land, and water are at least partial causes of cancer. Industrial agriculture is clearly responsible for some of this pollution.

It is so obvious that organic farming is better, yet society is caught in a heated debate regarding the merits of conventional and organic agriculture. Sadly, it is not a fair battle because there is so much money riding on the conventional, industrial side. Agribusiness corporations rely on the social acceptance of their synthetic fertilizers and pesticides. These businesses developed from World War II companies that manufactured chemical weapons. When this became unnecessary and socially unacceptable, they turned to warfare on agricultural lands with the development of chemicals to kill bugs and weeds. The infamous ones, like DDT and Agent Orange, clearly exemplify the dangers of using technology before we know the full consequences. There are others that have been banned, and there are probably many more agrichemicals that will be banned in the future.

But there is big money to be lost by conventional agribusiness if Americans finally realize the full impact of the ecological and social destruction wrought by the conventional agricultural system. It is killing our soil, polluting our air and water, using up our fossil fuels and our clean water, and harming our wildlife. At the same time, it is destroying our rural communities, decimating small towns, taking away farmers' independence, and pushing many family farmers into poverty. By focusing only on obtaining high yield and ignoring economic profit margins (the bottom line is that if farmers spent less on industrial inputs, they would have higher profits), the industrial agricultural system maintains itself because not enough people question it and seek change. We commonly hear the rallying call that American farmers "feed the world," but in fact they are feeding cows, pigs, and chickens. Seventy percent of corn grown in the United States is fed to livestock (USDA–NASS 2003). At the same time, we are told that we should reduce the amount of meat we eat! Why not shift some of the acres from feed-grain production to organic food crop production? Because even our diet is influenced by corporations.

As described in Nestle's book *Food Politics* (2002), our dietary guidelines are clearly influenced by the food industry, whose vast profits are used to buy off our politicians and scientific community. Initially, the recommended food pyramid was supposed to read, "Go easy on beverages and foods high in added sugars." But thanks to strong lobbying from the sugar industry, our guidelines now state, "Choose beverages and foods to moderate your intake of sugars" (Nestle 2002, 81). Whose advice is this, anyway? The food industry or independent nutritionists? Nestle notes that Americans drank twice as much soda pop as tap water in 1999, and the pop companies specifically target kids with their advertising. According to Nestle, we should organize at the grassroots level and protest the agenda of the food industry. We can learn lessons from the existing fight with Big Tobacco to formulate and effectively fight against Big Food. In addition to educating the population about "dietary literacy," one way to combat this questionable corporate influence on our government is simply to go back to fewer processed foods. Buy organic fruit, vegetables, and grains and cook them yourself.

Organic farming is a viable, ecologically responsible means of production that is increasingly acknowledged. Recipes in mainstream magazines, for example, now suggest using organic produce (e.g., Better Homes and Gardens 2003). At the same time there are a handful of irresponsible but much publicized foes of organic agriculture, who have made faulty claims about agriculture. Dennis Avery and his son Alex, both of whom are supported by the Hudson Institute, a conservative political group funded by agrichemical corporations such as Monsanto, Dow, and DuPont, are vocal opponents of organic agriculture (Avery 2000). They have cited inaccurate information in their arguments against organic farming. For example, Avery claims that the use of toxic natural pesticides such as dormant oils and sulfur has exploded because of organic farming. In fact, each of these is used more commonly on conventional farms, with dormant oils used on 70–90 percent of some industrial tree fruit crops and sulfur used on many other conventional fruit crops (Creamer 2001). Further, Dennis Avery interchanges the words *natural* and *organic* food when citing specific figures (Burros 1999), which any expert and most consumers know is inaccurate. Avery's vocal but invalid criticisms of organic farming are fictional at best.

Unfortunately, the Averys tend to make splashy stories for the media and are often portrayed as agricultural experts. Dennis Avery provided opinionated, unverified claims for John Stossel's inaccurate piece of journalism on the ABC television news show *20/20* in February 2000. Stossel told 12 million viewers that ABC commissioned laboratory tests of conventional and organic food. He claimed that organic food had about the same level of pesticide residues as the conventional food, which shows that it is a scam for people to pay more to buy organic food. After months of complaints, criticisms, and even an editorial in the *New York Times*, Stossel finally went on the air quietly to retract portions of his story in August 2000 (Rutenberg and Barringer 2000). Most notably, he admitted that the laboratory *never tested* for pesticides in the food samples, and that *the information he reported simply did not exist* (Burros 2002). But he did nothing to remedy the overall negative and inaccurate tone of his report. These are just a few examples of the misinformation from opponents of organic farming seeking to misrepresent it and the sensationalist reporters willing to fabricate a news story. It is important for advocates of organic agriculture to realize that naysayers exist, so we can vehemently discredit their assertions. We must expand the knowledge base and educate Americans about the true ecological and social potential of organic farming.

These corporate and media-induced injustices must be revealed. Agricultural and food issues should top our list of news stories in the United States, but unfortunately they are rarely mentioned. This is why strong advocacy can play a crucial role in bringing these issues to the forefront. Talk to your friends, your neighbors, your kids, your kids' teachers, and the school cafeteria cooks. Talk to the managers of your favorite restaurants and grocery stores. Tell them you want organic food to be available in your community and that locally grown organic food is even better. Be heard. Bring these issues to light.

RELEVANT RESEARCH

Organic farming research receives horrendously low levels of funding, compared with both our government's support and private industry's big bucks for conventional agricultural research. Increasing organic research money is an obvious need. Just imagine what we could learn from equitable research funding! Let's "allocate equal funding to research on nonproprietary approaches to agriculture (i.e., organic methods) as proprietary approaches (i.e., biotechnology) receive from their private sources, and then allow these two approaches to be packaged and 'marketed' to farmers and consumers" (Lotter 2003, 104). With this level playing field, even more producers would be attracted to organic methods and more consumers could be educated about the benefits of organic farming. Although equal funding may be a pipe dream, just a modest increase in the funding for organic research would provide valuable information that could increase yields, provide stable production, and help with distribution. While research funding priorities must be shifted, a deeper concern is how to make research truly relevant to organic farmers.

Even objective researchers do not always make the best decisions. For example, much research is a cautious reiteration of past studies. As plainly outlined in *Farming in Nature's Image* by Soule and Piper (1992), there are two ways to be wrong with statistical analysis. First is to claim there is an effect when in fact there is none. Second is to claim there is no effect when in fact there is one, but your methods did not detect it. "To say, 'I didn't detect an effect,' saves face better than 'I was fooled by the numbers'" (73). It is better to be safe and cautious than to go out on a limb and discover something remarkably new. "So scientists adjust statistics to allow only a small chance of the first type of error and a larger chance of the second. This bias protects the status quo" (73). This is particularly detrimental for organic farming, as it is outside of mainstream research and has been viewed for so long as nonviable and inconsequential. So if researchers only want to play it safe, they rarely conduct studies to discover the extraordinary benefits of organic methods, and they certainly would not want to find shocking new statistical data.

Organic agroecosystems are complex, which leads to another problem: science is most comfortable in reductionist approaches that study distinct, separate parts of a system. "Where chaos begins, science stops. For as long as the world has had physicists inquiring into the laws of nature, it has suffered a special ignorance about disorder in the atmosphere, in the turbulent sea, in the fluctuations of wildlife populations, in the oscillations of the heart and the brain. The irregular side of nature, the discontinuous and erratic side – these have been puzzles to science, or worse, monstrosities" (Gleick 1987, 3). Obviously, organic farming requires a holistic approach that is mostly outside the narrow disciplinary confines of industrial agricultural research. We must build research programs that are complementary: "There needs to be constant feedback between whole systems and components research" (van Bruggen and Termorshuizen 2003, 154).

In addition to the ecological interactions present on organic farms, economic factors are also embedded within the framework of tradition, family, and community. This is also not recognized within most agricultural economic research, "because the concept of sustainability is fundamentally incompatible with conventional economic theory." Indeed, "economics is a unidimensional, consumptive science. There is nothing within economic theory that reflects human values in terms other than people as consumers or as producers of goods for consumption. Society, the environment, and even preferences for non-consumption goods are considered as external to the economic decision making process." Clearly, economic research must draw from new ideas, and "the study of 'sustainability' will require a new, more inclusive theory of economics" (Ikerd 1997).

Likewise, researchers cannot assume they are the only experts in determining what topics of research to undertake; rather, we can learn a great deal by listening to farmers describe their on-farm management decisions, personal motivations, problems, goals, and experiences. This is a key concept of pragmatic approach, by which we seek to understand a complex situation through people's experience-based viewpoints. In the future, we must tailor research to specifically address farmers' needs for information about on-farm production, ecology, and marketing. The key issues in organic farming today, according to a 2002 workshop held by the Organisation for Economic Co-operation and Development, are sustainability (economic and environmental), markets (supply and demand), and policies (OECD 2002). These key topics should be melded into organic farmers' information demands and become the basis for active fieldwork, data collection, research analysis, and dissemination of results.

There is divergence between the topics of most agricultural research undertaken and the factors that organic farmers themselves indicate to be most influential in their farm operations. Even research that specifically focuses on organic farming has not always been on the most relevant topics. For example, some researchers theorize about the commodification of organic products within the agro-food complex, but they don't provide any real suggestions for how to slow these trends. Other research is on abstract economic modeling of consumer behavior, when in fact farmers seek specific information on marketing channels for selling their crops. Yet another example is seen in the technical field plot comparisons of conventional and organic crops for one crop season – organic farmers know that their methods work, but they really want long-term studies on the intricacies of soil fertility and pest management.

Farmers commonly note that research and information sources are not adequate, and much of the information available is not relevant to their farms. Since the typical sources of agricultural information (government and universities) have not been taking responsibility for this component of agriculture, other means have become more important. First, farmers conduct their own on-farm experiments to determine what organic methods work best in their local geographies. Programs to provide grant funds for this sort of research should be made widely available. Second, most organic farmers meet and share information, either formally at meetings of various organic farming organizations, or informally by calling or meeting individually with fellow organic farmers. Although some organic farmers are quite careful not to share marketing information, most are still willing to discuss on-farm production problems. In fact, most have learned to seek information far and wide and are willing to call anywhere in the world to obtain information on crop and livestock management.

In addition, a handful of groups have been established to disseminate information on organic farming. According to their Web site (ofrf.org), "the Organic Farming Research Foundation is a non-profit foundation founded to sponsor research related to organic farming practices, to disseminate research results to organic farmers and to growers interested in adopting organic production systems, [and] to educate the public and decision-makers about organic farming issues." They are, in fact, one of the few sources of research on purely organic farming methods, and they encourage other researchers and government agencies to give attention to organic farming through activities such as their Scientific Congress on Organic Agricultural Research (SCOAR).

The OFRF national farmer survey is the only source of national information on organic farmers and their farms. This survey, conducted in 1993, 1995, 1997, and 2001, is now sent to over 6,000 certified organic farmers nationwide and provides data on many aspects of their farms. The results of the 1997 survey are on-line, in a 130-page document that can be accessed easily and that provides a useful overview of the issues facing organic farmers today (Walz 1999). Results of various OFRF-funded research projects are also referred to on their Web site. It is likely that this sort of research and information will grow in importance, as organic farmers seek answers to their specific production questions. Several other international organizations provide information on organic agriculture, most notably the International Federation of Organic Agriculture Movements (IFOAM) and the Soil Association in the UK (see appendix).

Very recently, the USDA established the Integrated Organic Program through the Cooperative State Research, Education, and Extension Service (CSREES). Funds are available for research projects that "solve critical agriculture issues, priorities, or problems through the integration of research, education, and extension activities." This program seeks to "improve the competitiveness of organic producers" and help organic producers and processors "grow and market high quality organic food, feed, and fiber" (USDA–CSREES 2004). These goals are promising. The research projects funded through this program have the potential to be highly relevant to organic farmers and hopefully these research results will be disseminated in a manner that is readily available to them.

We are just beginning to see results of research that targets organic farming, but there remains much room for improvement. A specific research division should be set up by the USDA and each state department of agriculture that funds research on holistic agroecosystems research that is relevant to organic farmers. These programs must be linked directly to regional concerns, so that organic farmers are surveyed and asked to rank topics of importance. The key is to shift the focus from corporate research that tests agrichemical applications for profit, to public research that investigates real farm issues for the good of society.

PROMOTING ORGANIC METHODS

A handful of specific policy actions could create an enabling atmosphere for organic agriculture: establishing programs that pay farmers for their organic conservation actions, creating offices that provide technical assistance on organic methods, allowing voluntary increases in the small farm exemptions from organic certification, requiring country- or state-of-origin labels on all food sold in the United States, and protecting the meaning of National Organic Standards through the democratic process. Before discussing each of these policy options, we must first place U.S. organic agricultural policies within a broader context.

Drawing from European examples, Pugliese (2001) notes that organic farming has attained sociopolitical acceptance due to the commitment of early "pioneer" organic farmers, the current attraction of people from all social and demographic groups, and the emotional and sensory values toward food that organic farming evokes. To some extent, the American experience is similar. But in Europe, the status of organic farming puts it in a position to work toward policy changes in the rural countryside, whereas the United States is only in the initial stages of forming organic agricultural policies with the establishment of the National Organic Standards in 2002.

The United States should learn from the European example: strong governmental incentives can encourage and maintain organic agriculture (Padel et al. 1999; Lohr and Salomonsson 2000). U.S. government policy for organic farming needs to be built from the ground up. Subsidies should be provided for farmers who are converting to organic methods, since the three-year transition is a time of insecurity. During this time, organic premium prices are not yet available, new farming techniques must be learned, and the ecological components of the farm are just moving toward a new balance. Once an organic farm is established, the USDA should pay subsidies to maintain it. Rather than paying billions of dollars to support commodity production that causes environmental degradation and gluts the market to drive prices down, the government should reward organic environmental stewardship. A new federal program called Government Resources for Organic Work (GROW) should be established to pay farmers for their ecological (increased wildlife, plant diversity, integrated nutrient flows, and safe water) and social (rural development, community stability, and family farm sustainability) contributions to the rural countryside. Research indicates that an across-the-board payment per acre for organic farming contributions to our environment is clearly justified (O'Riordan and Cobb 2001), and more should be given for social benefits.

The main reasons why organic farmers are forced to quit are related to the inability to find successful markets and the lack of technical assistance for addressing in-field agricultural problems (Rigby et al. 2001). So government programs should be established to assist farmers in identifying reliable marketing options, especially regional marketing of organic food (e.g., assistance for farmers' markets, CSAs, farmers' cooperatives). Finally, farmers and the public deserve accurate, up-to-date information on organic farming.

One example of successful private efforts to increase both supply and demand of organic commodities was undertaken in the Upper Midwest Organic Marketing Project, with farmers from North Dakota, South Dakota, Minnesota, Iowa, and Wisconsin and consumers primarily from Minneapolis-St. Paul. Funded by a $1.25 million grant from the Pew Charitable Trusts, specific steps were taken to increase consumer knowledge about organic products (which was quite successful, even at large mainstream grocery stores) and to encourage farmers to adopt organic methods (which was somewhat less successful). Perhaps the most notable finding from this project is that "organic grain and soybean farming is best-suited to full-time farmers with moderate-sized operations" (Dobbs et al. 2000, 126) – like the family farms we've seen in previous chapters. And these family organic farmers "should seek to operate through effective marketing cooperatives" (128); farmers' cooperatives help keep more of the profits on-farm rather than lining the pockets of agribusinesses. Well-funded government-sponsored efforts could be based on this successful regional project and could stimulate nationwide increases in organic production and consumption – if the political will could be found to fund such efforts.

Recently, the USDA Agricultural Research Service started moving incrementally in the right direction (Jawson and Bull 2002), and USDA Economic Research Service researchers are conducting some studies on organic production. Still, the U.S. government should ensure that valid, consistent information on organic methods is available to farmers, and technical advice should be provided by the USDA, extension agents, and university researchers who focus solely on organic techniques.

Further, pending legislation that would require country-of-origin labels would help American farmers in general and organic farmers specifically. In fact, a "state-of-origin" label would be even more useful for consumers interested in trying to buy locally or regionally grown food. Unfortunately, agribusiness corporations find global markets lucrative, and they strongly oppose such labeling, as it informs consumers about how far their food is being shipped. But at the grocery store, we should have clear signs that indicate where our food is grown. Consumers who are informed about the issues can make informed decisions and try to buy foods produced regionally or nationally. Just as we hear "Buy American" in regard to industrial products, we should teach people about American farmers and how to support American agriculture. We must convince our politicians that an origin label benefits both farmers and consumers.

Caution should be exercised in regard to organic certification standards, now that the USDA has provided the culminating regulation. Using the Danish example – they are about fifteen years ahead of us – we can see what may be of concern in our near future. There was a notable change in Denmark, from farmers in control of organic standards to the national government's oversight. Farmers were gradually excluded from the process (Michelsen 2001a). So there have been major increases in organic food production and consumption in Denmark, and farmers are following good organic growing techniques, but they feel that the standards no longer represent their ideals and their social values. The United States is at the starting line on this issue. Already, many organic growers with smaller acreages are questioning the USDA's control of the term *organic*, since they could be fined up to $10,000 for using the word without being certified. There is a "small farm clause" that exempts farmers from this law if they have under $5,000 in annual sales, but this is a very low cutoff rate, applicable only to very small, part-time farms.

Thus farmers must decide if it is worth the price, the paperwork, and the time to be certified each year. Organic advocates should rally to increase this cutoff margin to at least $20,000 or $25,000 per year to allow small-scale local producers, who don't really "need" the certification label in order to sell their crops to local markets, to continue to call themselves organic without the cost of USDA certification. If local consumers trust the farmer, they don't need the certification label, and small producers cannot afford to pay for certification. Yet if these small-scale producers are using true organic methods, they clearly deserve to use the word *organic* (but not *certified organic*).

Overall, we must safeguard the integrity of organic certification, which has already been coming under attack. According to the *New York Times* (March 5, 2003), "If it weren't so dangerous, the chicken fight going on in Congress would be laughable." Incredibly, the very basis of organic standards, in place only a few months, was threatened by secretive Republican Deal making (pun intended). Representative Nathan Deal from Georgia received strong financial support from Fieldale Farms, a huge poultry operation headquartered in his state. Already in the spring of 2001, Deal and Fieldale were pressuring Secretary of Agriculture Ann Veneman to grant an exemption to the requirement that organic chickens be fed organically grown feeds (National Campaign for Sustainable Agriculture 2002). Claiming that organically grown grains were unavailable at prices that were acceptable to Fieldale, the Georgia delegation hammered away at Veneman. Luckily, she realized the importance of maintaining the integrity of organic certification (e.g., if a chicken eats pesticides, it's certainly not organic!), and she refused to grant this exemption. So Deal waited a few months and then simply slipped a paragraph into a $397 billion spending bill that indeed would allow farmers to give livestock nonorganic feed but still label their meat, eggs, and milk as organic (*New York Times*, March 5, 2003). Such a shortcut not only undermines organic livestock and their products but also hurts organic grain farmers who depend upon these sales. Quickly, organic proponents cried foul (and fowl) as consumers and food corporations both saw the hazards of this loophole. Members of both political parties jumped into action, as Senators Leahy and Snowe introduced legislation that would kill Deal's deal. And there was even talk of creating an Organic Caucus within Congress to protect the federal standards from other such assaults. We'll see how it goes. In the meantime, we must all be cautious and aware of legislation that pertains to organic farming. Staying informed is a key step in protecting certified organic farms. So keep your congressional telephone numbers handy and get on the phone to keep family organic agricultural issues on the legislative agenda.

Another consideration is what organic certification should mean. In the United States, it focuses exclusively on in-field production methods, with careful attention to the exclusion of prohibited substances. Certainly, this purely production-based standard of certification overlooks the deeper philosophical and ideological reasons behind organic farming, but the fact is that regulating a person's ideals is nearly impossible. On the other hand, U.S. certification could be greatly improved with the inclusion of more comprehensive ecological standards, such as on-farm biodiversity and landscape goals. These ecological benefits would be obvious at the farm level, as they require that land be set aside (not put into crops), which in turn indicates a farmer's commitment to noneconomic goals.

At present, the benefit of certified organic agriculture is its clear meaning. Farmers and consumers know what production methods are employed: no synthetic chemicals and no GMOs are used. As discovered in a comparison of eighteen European countries, common production standards are the most influential factor in increasing organic farming, since both farmers and consumers demand a clear and uniform definition (Michelsen 2001b). High certification standards and national regulations create consumer confidence, encourage genuine organic farmers, and provide easy access to international trade (Tate 1994). Of course, this globalization of organic food flies in the face of local food initiatives. This is a complex geographical issue, and we must strive to create a delicate balance between small organic farms that can supply nearby communities and larger organic farms that can produce grains for national and international distribution (while encouraging other nations to produce their own organic crops). We should support the existing USDA standards but at the same time work to strengthen them to benefit family organic farmers, the environment, local communities, and consumers alike. This is the best way to promote the continued success of U.S. organic farming.

PROTECTING ORGANIC FARMING AS AN ALTERNATIVE

The success of organic farming is becoming clear, and Americans are beginning to see that this form of agriculture can provide ecological and social benefits. At the same time, this mainstreaming has led to another problem: agribusiness (or Big O Ag) realizes that organic farming is a profitable segment of agriculture. So we must promote organic farming in a form that can "coexist with the global industrial food system rather than being co-opted by it" (Milestad and Darnhofer 2003, 94). And indeed, one corporation already controls many of the familiar organic product lines we see in the stores: the Hain-Celestrial Group (of which Heinz has 20 percent equity) owns Earth's Best baby food, Nile Spice, Garden of Eatin', Arrowhead Mills, Health Valley, Casbah, Imagine/Soy Dream, Celestial Seasonings, Westbrae, Westsoy, Little Bear, Bearitos, etc. (Howard 2004). The familiar organic labels Cascadian Farm and Muir Glen have been owned by General Mills since 2000 (Halweil 2001; Sligh and Christman 2003). The product labels carefully omit listing this corporate giant, and according to a *Wall Street Journal* article, General Mills is well aware of the importance of organic brand identity (Helliker 2002). A marketing report called Supermarket Strategic Alert (2003, 2) notes that large corporations realize how natural food shoppers "are leery of the national brands" and "the parent company may actually be a negative, which is why it may not be named on the package." Along these same lines, organic milk and soy beverages are following the conventional path of concentration, as Horizon Organic Dairy now controls about 70 percent of organic milk distribution (Brewster 2002). In early 2004, Horizon was acquired by Dean Foods, the largest fluid milk producer in the United States, which already owns White Wave (Silk and Sun Soy) products (Sligh and Christman 2003; Standard and Poor's 2003). To continue with my earlier recommendation for product labeling: we should push for "truth in corporate ownership" legislation so that food companies must disclose the corporate giants that own many organic brands. Such labeling would help consumers make informed choices in both the supermarket and the natural foods store.

Much of the research that discusses the problem of Big O Ag is from California, where organic farming is more accepted and where there is much greater availability of organic foods. This contrasts with the U.S. Midwest, where organic farming is still viewed as a fledgling alternative approach that needs to be protected by organic farming advocates. For example, when I visited a friend in northern California, she took me to an average neighborhood grocery store that was stocked with four kinds of organic apples in bulk and multiple other organic fruits and vegetables, plus a full array of organic dairy and meat products, and a huge variety of organic packaged foods. I was amazed! I live in a university town in southern Illinois (population 27,000 plus 20,000 students). In our biggest, most "upscale" supermarket there *might* be a few three-pound bags of organic apples, a few heads of organic broccoli, and a few two-pound bags of organic carrots (but not always, and this has only occurred within the past few years). There is also a small section labeled "Specialty Foods," but this is not all organic; much of it is "natural" food, whatever that means. Luckily, we also have an excellent cooperative grocery store in town that sells a variety of organic items, but this is not frequented by "mainstream" shoppers whose only experience with organic products is seeing those few items at the local supermarket. The supermarkets in most smaller towns or nonuniversity towns in the Midwest and South stock zero organic products. Research indicates that consumers at supermarkets are less likely to accept information on organic agricultural issues than shoppers at co-ops and farmers' markets who are more open to the inclusion of educational materials as part of the shopping experience (Schäfer 2003). There is substantial geographical variation in the availability of and appreciation for organic items.

This issue of "conventionalization" in which organic farming becomes dominated by Big O Ag agribusiness is a very real concern, but it hasn't happened in Europe, and they are about a generation ahead of us in the development of organic farming. We must learn from their example. Take Denmark: they created federal organic regulations in 1987 and have implemented many national regulations that encourage organic production since then. In terms of marketing and the role of agribusiness, there are interesting variations within the Danish situation, as small-scale organic farmers have worked within the conventional system to sell their products in supermarkets and yet remain independent and part of distinct organic farming groups (Michelsen 1996). We can hope that this will happen here, and we can do something about it – to ensure that organic represents a true "alternative" in the United States, well into the future.

The simplest action for an organic advocate is to buy from local organic farmers. Know your farmers. The more you know about the farmer who produces the food you eat, the better assured you are that the methods she or he uses protect the natural environment of your local region. Buying organic food from a local farmer puts money right back into your local economy, and this helps maintain a healthy rural landscape. There are increasing opportunities for these regional, organic food purchases; CSAs are being established across the country, farmers' markets are booming, and many food co-ops stock locally grown produce. These types of local activities could be part of the "emerging 'unique' markets" that will pose a challenge for the global agricultural corporations (Hendrickson and Heffernan 2002b). These ideas are echoed in the book *Slow Food* (Petrini 2001), which promotes a movement to counter the fast-food culture that is threatening to take over the world. It urges us to "operate within a regional framework and promote new forms of 'slow' production and supply" because this guarantees quality food and "pays due respect to agriculture" (2).

The Farm as Natural Habitat (Jackson and Jackson 2002) describes the "connection between the grocery list and the endangered species list, between farming and nature" (2). The authors make a strong case for buying regionally to support farmers who are practicing ecological methods of agriculture. While they do not endorse certified organic agriculture as a whole (noting that agribusiness influences are increasing), they support the activities that true family organic farmers undertake: ecological integrity and soil building through crop rotation and farm diversification. Alternative methods that promote healthy ecosystems and biodiversity right on the farm is the solution to the problems of industrial agriculture. The best way to promote this type of farming is to be aware of the impacts of farming and make agriculture a key *environmental* issue. Ideally consumers should have the opportunity to buy from local organic farmers whom they know and trust.

The geographic reality is that we cannot always buy local organic food. Climatic conditions and seasonal variations determine what food can be grown in what regions. And while grassroots organic growers and activists may oppose them, the bigger and less self-reliant organic farms also have a place in organic agriculture – as long as these larger organic farms still provide a real opportunity for industrial American farmers to convert to organic methods. Bigger organic farms should provide a means to keep a farm family on the land. Granted, this is not the small-scale, local, self-reliant organic farms we idealize, but it could still be a valuable component of a sustainable rural countryside. These larger organic farms would still use holistic organic techniques, not conventional chemical controls. So they are different from their industrial neighbors. Certainly locally marketed, small-scale organic farms are a step away from the industrial agricultural system, but a huge leap occurs when a medium or large-scale conventional farm family is able to make the transition to organic methods. We need the small-scale organic farms supplying a variety of vegetables to local and regional markets, and we also need the midsized, family organic farms selling to local, regional, national, or even global markets that represent a wholesale shift away from industrial production. Consumers must take responsibility to be sure that these family organic farms are able to remain independent and viable; we need to ensure that corporate Big O Ag doesn't besiege them.

To help consumers identify true family-operated organic farms, we could create a new certification label, called Fair Share (Brussell 2003). When used in combination, the "Certified Organic and Fair Share" label would verify that the products were grown by family farmers who earn a fair price (say, at least 75 percent of consumer price) for the products sold, and that it is marketed outside the grips of agribusiness corporate control. For a consumer this is not as direct as buying from the local farmers, but it would still provide linkages from the field to the table. In addition, such labels could bolster farmer cooperatives through which farmers join together to market their commodities and gain a higher profit. Organic Valley Family of Farms is just one example of a thriving farmer cooperative that markets nationwide (see organicvalley.com).

American farmers have to confront both old and new ideas about farming. If they grew up on a farm, they likely remember the post–World War II era of chemical production and may find it difficult to find any alternatives. On the other hand, most farmers know the dangers of agrichemicals, are leery of the chemicals, and sometimes seek a way out of this type of production. Family tradition can be a stumbling block for farmers seeking new methods. "We have farmed this land with conventional methods for generations, so why should I do it differently?" Plus they feel this is the only way to obtain the high yields demanded in the industrial system of agriculture. This is all they know. This is the only system they've seen.

We cannot expect to take hundreds of 1,000-acre industrial corn farms and turn them each into 100 ten-acre organic vegetable farms – the markets, consumers, farmers, and rural areas are simply not ready for that. But we could try to take the 1,000-acre farms and break them into four organic farms of 250 acres each that are diverse crop and livestock operations. This is realistic now. Then, in the future, we could work toward changing the agricultural system, which is based on meat dependence. If we reduce the need for feed grains, we would need more, smaller organic vegetable/grain/legume farms. We must think big, but take appropriate, realistic steps now to change our agriculture. We must make pragmatic changes that can actually happen sooner, rather than make theoretical plans that can only – if ever – be implemented much later. We need to support midsized family organic farms that can lead us in the direction of a complete shift to organic methods.

OUR ORGANIC FUTURE

Organic farming has proven itself as a viable option for farmers seeking to work outside the industrial agricultural system. It is no longer a question of whether organic farms can survive economically or produce enough food to sustain us. Now we must ask the more complex and thought-provoking questions: what level of ecological diversity can an organic farm achieve, and how many distinct marketing channels can an organic farmer develop?

We know that organic agriculture is best for the environment, farmers, and society. Yet we are torn between two visions of organic farming: the historical grassroots organic movement was based on small integrated organic farms that sell locally versus the recent but increasing agribusiness interests (Big O Ag) that have the monetary and political influence to win out and possibly dominate organics. In the middle are midsized family organic farms, which I profiled in chapters 4 and 5. These farmers are the real heroes, the real future of organic agriculture – if they can overcome both the grassroots critics and the powerful forces of agribusiness conquest.

Organic farming is pulled in these two directions because of its success. Yet, even with the growing acceptance of organic farming, our relationships toward rural America are still complex and unresolved. Urbanites imagine a romantic rural countryside with farms dotting the landscape and the rural folks living in harmony with the environment, but in truth the industrial agricultural setting is very different. Many industrial operations are more like a factory located on an exploitable piece of land, rather than a farm integrated into its ecological and social surroundings. And many of the rural people in this setting also relate to agriculture as if it were merely an industry rather than a primary activity in which people work closely with the land.

This became clear to me in a small diner on the plains of eastern Colorado. This is industrial ag country, with average farm size at several thousand acres. I ordered a cup of coffee, and here the choice is coffee or decaf (forget about fancy urban choices like cappuccino). As the waitress sloshed the plain white ceramic mug down in front of me, she reached in her apron to toss me a few packets of non-dairy creamer. (This stuff is so chemical-laden it is actually flammable – test it sometime.) I politely asked for some milk to put in my coffee instead, and the waitress looked at me as if I were from Mars. "Why?" she asked incredulously. "I just like it better," I meekly responded. But I wanted to shout, "I don't want to put mysterious chemicals in my coffee! You are here in rural America. You should understand the value of fresh air and nature. If the farms here weren't industrialized, you would be able to see the links between rural life and nature." But most areas of industrial agriculture are indeed this removed from their regional geography. Farmers sit in the air-conditioned cabs of their tractors and drive their thousand-acre fields, spraying pesticides or chemical fertilizers (or paying the agrichemical dealer to do it). They are removed from the nature that should actually grow their crops, because it is mostly suppressed by technology. And economically they are hardly able to make a living on the land unless they control thousands of acres. All the while, many urban Americans decorate their kitchens with country motifs of red barns and sing "Old MacDonald" to their kids. But Farmer MacDonald is hard to find nowadays – with one cow, one pig, a rooster, and a chicken, E-I-E-I-O. More likely the industrial farmer has a confined feeding operation of three thousand hogs and has to spray tons of toxic chemicals in order to control weeds on his thousand-plus acres of corn . . . all in an attempt to scrape out a living in modern industrial agriculture. Clearly we need an alternative.

Certified organic farming is a viable alternative. Organic farming is the best alternative to the many problems of current U.S. industrial agriculture. While smaller-scale organic vegetable farms have already made an important mark on U.S. agriculture, medium to large family organic farms must also flourish. Even at this larger scale, organic farmers remain inherently closer to nature, as they must be in their fields scouting for insects and checking soil fertility and deciding the next crop rotation. This book has shown that different types of organic farms in various geographic settings are successful. With perseverance and courage, organic farmers can face industrial agriculture head-on and win. Organic farming provides a distinct, definable method of production that gives farmers more options for selling their crops and the opportunity for increased connections with consumers. If organic farms are accessible, people will be encouraged to learn a bit more about farming – where and how their food is produced. And farmers should have more flexibility to diversify and earn a fair wage for these organic crops and livestock. Certified family organic farms can help create a socially vibrant and sustainable rural countryside within a landscape that promotes the ecological integrity of our water, air, and soil. This should be the rural geography of our future. Let's make it a reality.

Appendix

Information Links

INTERNATIONAL ORGANIZATIONS

Canadian Organic Growers (http://www.cog.ca) is a membership-based information network that represents farmers, gardeners, and consumers across Canada. COG has been active since 1975 and now participates in conferences, publishes a quarterly newsletter, and has a free mail service organic library. They promote organic agriculture and the environmental, health, and social benefits that go with it. Their *Organic Fieldcrop Handbook* and *Organic Livestock Handbook* are excellent sources.

The European Union Commission Agriculture Directorate-General set up a Web site (http://www.organic-europe.net) with the help of the Germans (Stiftung Ökologie und Landbau) and the Swiss (Forschungsinstitut für biologischen Landbau). This Web site provides information on organic farming in twenty-five European countries.

The European Environment Agency provides information to policymakers so they can make sound policies to protect the environment and support sustainable development in the European Union (http://www.eea.eu.int). A report entitled "Organic Agricultural Research in Europe" is found at http://ewindows.eu.org/Agriculture/organic/Europe/Report.

The Food and Agriculture Organization of the United Nations provides organic agriculture information at http://www.fao.org/organicag, including FAO documents, international contact information, agricultural data, and meeting announcements.

The International Federation of Organic Agriculture Movements has the status of "consultant" to the UN and FAO. IFOAM is an umbrella organization for nearly 750 organic agricultural groups, representing one hundred countries. It promotes the global application of organic methods and provides avenues for information exchange through conferences and publications. The IFOAM Web site (http://www.ifoam.org) provides links to affiliated organizations.

The Organisation for Economic Co-Operation and Development (OECD) is an international group whose mission is to help governments "tackle the economic, social, and governance challenges of a globalised economy." The OECD Agricultural Directorate specifically deals with food, agriculture, and fishery issues and seeks to provide information governments on practical and innovative options for the reform and development of policies and the liberalization of trade. They have sponsored several workshops and reports on organic farming, which are linked within their Web site (www.oecd.org).

The focus of the Soil Association is the link between healthy soil, healthy food, and healthy people. The Soil Association, founded in 1946, is the main organic certification organization in Great Britain, active in promoting organic food and farming. It relies on donations from members, fees from certification, and substantial grants from the British government. Its Web site (http://www.soilassociation.org) has sections on standards and certification, farming and growing, and manufacturing and retailing, to name a few.

ORGANIC FARMING ORGANIZATIONS BASED IN THE UNITED STATES

Beyond Pesticides (formerly the National Coalition Against the Misuse of Pesticides) is a national organization that seeks to reduce or eliminate toxic pesticides. It educates about pesticide safety and alternative pest management techniques. See http://www.beyondpesticides.org.

The Campaign to Label Genetically Engineered Foods has been active since 1999. Its Web site (http://www.thecampaign.org) provides action alerts (suggestions for contacting food corporations, congressional leaders, or other politicians), updates on genetically engineered crops and related legislation, and educational information.

The International Center for Technology Assessment (http://www.icta.org) is a Washington DC–based nonprofit organization that provides information on the economic, ethical, social, environmental, and political impacts of technology. One of their projects is the Center for Food Safety (http://www.centerforfoodsafety.org), which focuses on food, environmental, and agricultural issues. Specifically, the CFS works on organic food issues and seeks labeling of genetically engineered foods. They take legal action to advocate for these issues.

The Environmental Working Group is a nonprofit environmental research organization funded by grants from foundations, with help from individual donors. It conducts computer-assisted research on environmental issues, including pesticides in foods, air, and water. According to its Web site (http://www.ewg.org), "The goal of EWG's research is to turn raw data into usable information." Check out their listings of foods with high pesticide residues and their posting of government subsidies to agriculture.

The Institute for Food and Development Policy, better known as Food First, is a member-supported, nonprofit think tank and education-for-action center. Food First was founded in 1975 by Frances Moore Lappé and Joseph Collins, following the international success of the book *Diet for a Small Planet*. The Food First Information and Action Network (http://www.foodfirst.org) is its action and campaigning partner whose work highlights root causes and value-based solutions to hunger and poverty around the world, with a commitment to establishing food as a fundamental human right. It provides an informative overview of Cuba's transition to organic farming.

Greenpeace can be found at http://www.greenpeace.org/homepage. Founded in 1971, Greenpeace is now an international organization in forty countries across Europe, the Americas, Asia, and the Pacific. In order to remain independent, Greenpeace relies solely on individual supporters and foundation grants; it does not accept donations from governments or corporations. It has numerous action issues, including elimination of toxic chemicals and rejection of genetic engineering. The True Food Network (http://www.truefoodnow.org) is a site sponsored by Greenpeace to address the problems of genetically engineered foods.

Growing for Market is a national on-line monthly newsletter about growing and marketing produce, herbs, and cut flowers. It is aimed at farmers who market directly to consumers. Unlike most farming magazines, every article is written by an experienced farmer. Subscription information is found on-line at http://www.growingformarket.com.

As part of the international Penton Media, Inc., New Hope Natural Media (http://www.newhope.com) publishes fifty-four business magazines and produces fifty trade show events throughout the world. New Hope works with natural products companies to reach their markets through print and on-line. The organic component (http://www.newhopeorganics.com) informs consumers and businesses about organic products, holds natural products expos, and publishes the *Natural Foods Merchandiser*, which contains data on organic production and consumption.

The Organic Consumers Association Web site (http://www.organicconsumers.org) states: "The Organic Consumers Association is a public interest organization dedicated to building a healthy, safe, and sustainable system of food production and consumption. We are a global clearinghouse for information and grassroots technical assistance. We currently have 500,000 people in our data base, including subscribers to our electronic newsletter, members, volunteers, and supporters, and 1,800 cooperating retail co-ops, natural food stores, CSAs, and farmers' markets. We have 1,000 key volunteers and coordinators working on developing OCA/BioDemocracy action teams across the country."

The Organic Farming Research Foundation (http://www.ofrf.org) is a nonprofit foundation with a very practical agenda: it sponsors research related to organic farming and disseminates the research results to organic farmers and those who may have adopted organic methods. Its informative publications include the National Organic Farmers' Survey, information on the lack of publicly funded organic research, and a newsletter.

Since 1985, the Organic Trade Association (http://www.ota.com) has promoted organic products and policy in Canada, the United States, and Mexico. Its members include growers, shippers, processors, certifiers, farmer associations, brokers, manufacturers, consultants, distributors, and retailers.

The Pesticide Action Network–North America (http://www.panna.org) is one of five PAN regional centers worldwide (the others are in Africa, Asia, Europe, and Latin America). By linking people at the local, national, and international levels, PANNA seeks to end the global reliance on pesticides and develop ecologically sound and socially just alternatives. It networks with consumer, labor, health, environment, and agriculture groups to build an international citizens' action network.

The Robyn Van En Center for Community Supported Agriculture (http://www.csacenter.org) provides a great deal of information on CSAs, including a national listing of active farms. Its focus is the northeastern United States for information on technical assistance, but general links on CSAs are useful, regardless of location.

The Rodale Institute was founded in the 1930s when J. I. Rodale began farming sustainably in rural Pennsylvania, with the notion that the key to healthy crops was healthy soil. Rodale, his son, and his grandson built what is now a world-renowned research center and publisher for organic, health-related information. Their Web site (http://www.rodaleinstitute.org) provides information on their field trials, magazines, and other resources. *Organic Gardening* has become an important magazine for small-scale organic producers, and *The New Farm* is an up-and-coming magazine with a great deal of potential. *Organic Style* is their trendy magazine, which deals secondarily with agriculture.

The Union of Concerned Scientists (http://www.ucsusa.org) includes 60,000 concerned citizens and scientists across the country. It seeks to protect the environment by joining scientific analysis and citizen advocacy. UCS works to be a "powerful voice for change" with interest in the risks of genetically engineered crops and the misuse of antibiotics in livestock.

ORGANIC FARMING EXCHANGE ORGANIZATIONS

Those people seeking "Educational Exchanges in Sustainability" can go to http://www.organicvolunteers.com. The Web site allows farmers and volunteers to sign up with the hope of "finding" each other and making a good match. The details of the volunteering arrangements are worked out by the hosts and volunteers themselves.

World Wide Opportunities on Organic Farms (http://www.wwoof.org) is an international cultural exchange program. The WWOOFers, as they are called, come from numerous nations and work on farms in many diverse locations: Australia, Denmark, Ghana, and Korea, for example.

Through the Multinational Exchange for Sustainable Agriculture (http://www.mesaprogram.org/index.html), international participants come to work and learn on U.S. organic farms. Each year approximately thirty individuals receive intensive organic farm training, free lodging, meals, and a small monthly stipend from participating U.S. host farms.

ORGANIC INFORMATION FROM THE U.S. GOVERNMENT

Most of the information on organic agriculture that comes from the U.S. government is framed within an economic context. It seems the federal government believes it is acceptable to provide basic information on organic agriculture because some farmers, wholesalers, and distributors are making money from it. But no other aspects of organic production are emphasized; U.S. governmental documents provide little discussion of the potential benefits to individual farm families, rural communities, landscapes, or environments.

Within the Environmental Protection Agency section on pesticides, some information on organic food is provided at http://www.epa.gov/pesticides/food/organics.htm. The Food and Drug Administration has a division called the Center for Food Safety and Applied Nutrition, which offers information on pesticides in food at http://vm.cfsan.fda.gov.

The National Medical Library's PubMed is a service of the National Library of Medicine that provides access to over 12 million MEDLINE citations back to the mid-1960s and additional life science journals. Its Web site (http://www.ncbi.nlm.nih.gov/entrez/query.fcgi?db=PubMed) offers links to full text articles and other related resources.

UNITED STATES DEPARTMENT OF AGRICULTURE AGENCIES

The National Agricultural Library (http://www.nal.usda.gov) is one of four national libraries (the others are the Library of Congress, the National Library of Education, and the National Library of Medicine). Its Alternative Farming Systems Information Center (AFSIC) focuses on locating information on alternative agriculture. Its three main areas are organic food production, sustainable agriculture, and community-supported agriculture (http://www.nalusda.gov/afsic).

The USDA–Agricultural Marketing Service (AMS) is in charge of programs that help Americans efficiently market their agricultural products (http://www.ams.usda.gov). It claims to "promote a strategic marketing perspective that adapts product and marketing practices and technologies to the issues of today and the challenges of tomorrow."

The USDA–AMS oversees the National Organic Program (http://www.ams.usda.gov/nop). This Web site lists certifying agents, consumer information, NOP regulations and policies, information for producers, processors, and handlers. In addition, the NOP provides a good overview of the current regulatory issues related to organic agriculture in their Today's News and the News Room. Its link to the National Organic Standard's Board (http://www.ams.usda.gov/nosb) describes this group and its delicate advisory status. It voluntarily advises the NOP, but the NOP is not mandated to take NOSB's advice!

USDA–AMS Farmers' Market Web site at http://www.ams.usda.gov/farmersmarkets provides facts about farmers' markets, including the National Farmers' Market Directory, which lists them by state.

The Appropriate Technology Transfer for Rural Areas (ATTRA) is a "national sustainable agriculture information service" managed by the National Center for Appropriate Technology and funded by the USDA Rural Business–Cooperative Service. Since 1987, ATTRA has sought to provide assistance to farmers, ranchers, extension agents, educators, and others interested in U.S. agriculture. Its Web site is http://attra.ncat.org.

The USDA–Economic Research Service (ERS) claims to promote the following components of American agriculture: competitiveness, food safety, human health, the environment, and rural quality of life. Given our current problems within each of these categories, the ERS has its work cut out. They have recently accepted the presence of organic agriculture and published several good reports that detail current trends in organic consumption and production. See the Organic Briefing Room at http://www.ers.usda.gov/briefing/Organic for links to data and information on organic agriculture.

Within the USDA's Food Safety and Inspection Service (FSIS) is the Office of Policy, Program Development and Evaluation, Labeling and Consumer

Protection Staff, which works with the National Organic Program on some labeling issues (http://www.fsis.usda.gov).

The USDA Foreign Agricultural Service (FAS) publishes attaché reports, one of which describes organic agricultural issues in specific countries. Through its Horticultural and Tropical Products division, its Organic Perspectives Newsletter describes international marketing prospects for organic goods (http://www.fas.usda.gov). Its Hot Markets section lists current trends in the demand for organic products.

USDA's Sustainable Agriculture Research and Education (SARE) administers research and education grants, producer grants, and professional development grants, and its related information branch, Sustainable Agriculture Network (SAN), publishes agricultural reports on its Web site (http://www.sare.org). Some of these programs and publications are related to organic methods.

References

ABC News. 2001. Poll: Behind the Label: Many Skeptical of Bio-Engineered Food, Organic Advantage. June 19. http://abcnews.go.com/sections/scitech/DailyNews/poll010619.html (accessed January 6, 2004).

———. 2003. Genetic Resistance. Poll: Modified Foods Give Consumers Pause. July 15. http://abcnews.go.com/sections/business/Living/poll030715_modifiedfood.html (accessed January 6, 2004).

Alavanja, M. C., D. P. Sandler, C. J. McDonnell, C. F. Lynch, M. Pennybacker, S. H. Zahm, J. Lubin, D. Mage, W. C. Steen, W. Wintersteen, and A. Blair. 1998. Factors Associated with Self-Reported, Pesticide-Related Visits to Health Care Providers in the Agricultural Health Study. *Environmental Health Perspectives* 106(7): 415–20.

Allen, Patricia, and Martin Kovach. 2000. The Capitalist Composition of Organic: The Potential of Markets in Fulfilling the Promise of Organic Agriculture. *Agriculture and Human Values* 17(3): 221–32.

Altieri, Miguel. 1987. *Agroecology*. 2d ed. Boulder CO: Westview Press.

———. 1999. The Ecological Role of Biodiversity in Agroecosystems. *Agriculture, Ecosystems and Environment* 74(1–3): 19–31.

———. 2001. *Genetic Engineering in Agriculture: The Myths, Environmental Risks, and Alternatives*. Oakland CA: Food First.

Andreoli, M., and V. Tellarini. 2000. Farm Sustainability Evaluation: Methodology and Practice. *Agriculture, Ecosystems and Environment* 77(1–2): 43–52.

Asami, Danny, Y-J. Hong, D. Barrett, and A. Mitchell. 2003. Comparison of the Total Phenolic and Ascorbic Acid Content of Freeze-Dried and Air-Dried Marionberry, Strawberry, and Corn Grown Using Conventional, Organic, and Sustainable Agricultural Practices. *Journal of Agriculture and Food Chemistry* 51: 1237–41.

ATSDR. 2002. U.S. Department of Health and Human Services Agency for Toxic Substances and Disease Registry. ToxFAQs for Aldrin and Dieldren. http://www.atsdr.cdc.gov/tfacts1.html (accessed January 6, 2004).

Aude, E., K. Tybirk, and M. B. Pedersen. 2003. Vegetation Diversity of Conventional and Organic Hedgerows in Denmark. *Agriculture, Ecosystems and Environment* 99(1–3): 135–47.

Avery, Dennis T. 2000. *Saving the Planet with Pesticides and Plastic: The Environmental Triumph of High-Yield Farming.* 2d ed. Washington DC: Hudson Institute.

Baker, Linda. 2002. The Not-So-Sweet Success of Organic Farming. *Salon.com*. Technology and Business. July 29. http://archive.salon.com/tech/feature/2002/07/29/organic/ (accessed January 6, 2004).

Barrett, H. R., A. Browne, P. Harris, and K. Cadoret. 2002. Organic Certification and the UK Market: Organic Imports from Developing Countries. *Food Policy* 27: 301–18.

Batte, Marvin T., D. Lynn Foster, and Fred J. Hitzhusen. 1993. Organic Agriculture in Ohio: An Economic Perspective. *Journal of Production Agriculture* 6(4): 465–542.

Beingessner, Paul. 2003. Monsanto Sues and Sues and Sues and . . . *Crop Choice: An Alternative News Source for American Farmers.* July 14, 2003. http://www.cropchoice.com/leadstry.asp?recid=1855 (accessed January 6, 2004).

Berry, Wendell. 1977. *The Unsettling of America: Culture and Agriculture*. San Francisco: Sierra Club Books.

———. 1998. In Distrust of Movements. Speech for the Tri-State Environmental Educators' Workshop, Evansville, Ind.

Better Homes and Gardens Recipe Center. 2003. Golden Mashed Potatoes with Leeks and Sour Cream. http://www.bhg.com/bhg/recipe (accessed October 28, 2003).

Beus, Curtis E., and Riley E. Dunlap. 1990. Conventional versus Alternative Agriculture: The Paradigmatic Roots of the Debate. *Rural Sociology* 55(4): 590–616.

———. 1992. The Alternative-Conventional Agriculture Debate: Where Do Agricultural Faculty Stand? *Rural Sociology* 57(3): 363–80.

———. 1994. Agricultural Paradigms and the Practice of Agriculture. *Rural Sociology* 59(4): 620–35.

Björklund, Johanna, Karin E. Limburg, and Rydberg Torbjorn. 1999. Impact of Production Intensity on the Ability of the Agricultural Landscape to Generate Ecosystem Services: An Example from Sweden. *Ecological Economics* 29: 269–91.

Blackburn, James, and Wallace Arthur. 2001. Comparative Abundance of Centipedes on Organic and Conventional Farms, and Its Possible Relation to Declines in Farmland Bird Populations. *Basic and Applied Ecology* 2(4): 373–81.

Blomquist, Joel, Janet Denis, James Cowles, James Hetrick, R. David Jones, and Norman Birchfield. 2001. Pesticides in Selected Water-Supply Reservoirs and Finished Drinking Water, 1999–2000: Summary of Results from a Pilot Monitoring Program. USGS and USEPA Report 01–456. http://md.water.usgs.gov/nawqa/OFR_01–456.pdf (accessed January 6, 2004).

Brandt, Kristen, and Jens Peter Mølgaard. 2001. Organic Agriculture: Does It Enhance or Reduce the Nutritional Value of Plant Foods? *Journal of the Science of Food and Agriculture* 81: 924–31.

Brewster, Elizabeth. 2002. Ready to Rule: Organic Beverages Expect Boost from New Labeling Requirements. *Beverage Industry* 93(3): 61–66.

British Medical Association. 1999. *The Impact of Genetic Modification on Agriculture, Food, and Health*. London: BMA.

Brock J. W., L. J. Melnyk, S. P. Caudill, L. L. Needham, and A. E. Bond. 1998. Serum Levels of Several Organochlorine Pesticides in Farmers Correspond with Dietary Exposure and Local Use History. *Toxicology and Industrial Health* 14(1–2): 275–89.

Brown, Allison. 2002. Farmers' Market Research, 1940–2000: An Inventory and Review. *American Journal of Alternative Agriculture* 17(4): 167–76.

Browne, A. W., P. Harris, A. Hofny-Collins, N. Pasiecznik, and R. Wallace. 2000. Organic Production and Ethical Trade: Definition, Practice, and Links. *Food Policy* 25: 69–89.

Brussell, Juli, program director, Illinois Stewardship Alliance, Community Food and Farming Systems Program. 2003. Personal communication. October 10.

Buchmann, Stephen L., and Gary Paul Nabhan. 1996. *The Forgotten Pollinators*. Covelo CA: Island Press.

Buck, Daniel, Christina Getz, and Julie Guthman. 1997. From Farm to Table: The Organic Vegetable Commodity Chain of Northern California. *Sociologia Ruralis* 37(1): 3–20.

Burros, Marian. 1999. Anti-Organic and Flawed. Eating Well, *New York Times*, February 17.

———. 2002. Study Finds Far Less Pesticide Residue on Organic Produce. *New York Times*, May 8.

Butler, Leslie J. 2002. The Economics of Organic Milk Production in California: A Comparison with Conventional Costs. *American Journal of Alternative Agriculture* 17(2): 83–91.

Buttel, Frederick H., and Gilbert W. Gillespie Jr. 1988. Preferences for Crop Production Practices among Conventional and Alternative Farmers. *American Journal of Alternative Agriculture* 3(1): 11–17.

Byrne, Patrick J., Richard Bacon, and Ulrich Toensmeyer. 1994. Pesticide Residue Concerns and Shopping Location Likelihood. *Agribusiness* 10(6): 491–501.

Byrum, Allison. 2003. Organically Grown Foods Higher in Cancer-Fighting Chemi-

cals than Conventionally Grown Foods. American Chemical Society public release. March 3, 2003. http://www.eurekalert.org/pub_releases/2003–03/acs-ogf030303.php (accessed January 6, 2004).
Cacek, Terry, and Linda L. Langner. 1986. The Economic Implications of Organic Farming. *American Journal of Alternative Agriculture* 1(1): 25–29.
Campbell, Hugh, and Ruth Liepins. 2001. Naming Organics: Understanding Organic Standards in New Zealand as a Discursive Field. *Sociologia Ruralis* 41(1): 22–39.
Canadian Broadcasting Corporation. 2003. Blowin' in the Wind. National Online. http://www.tv.cbc.ca/national/pgminfo/canola/ (accessed January 6, 2004).
Carson, Rachel. 1962. *Silent Spring*. Boston: Houghton Mifflin.
Cavieres, M., J. Jaeger, and W. P. Porter. 2002. Developmental Toxicity of a Commercial Herbicide Mixture in Mice. I. Effects on Embryo Implantation and Litter Size. *Environmental Health Perspectives* 110: 1081–85.
Chamberlain, D. E., J. D. Wilson, and R. J. Fuller. 1999. A Comparison of Bird Populations on Organic and Conventional Farm Systems in Southern Britain. *Biological Conservation* 88: 307–20.
Clark, Sean, Karen Klonsky, Peter Livingston, and Steven Temple. 1999. Crop-Yield and Economic Comparisons of Organic, Low-Input, and Conventional Farming Systems in California's Sacramento Valley. *American Journal of Alternative Agriculture* 14(3): 109–21.
Clemetsen, Morten, and Jim van Laar. 2000. The Contribution of Organic Agriculture to Landscape Quality in the Sogn og Fjordane Region of Western Norway. *Agriculture, Ecosystems and Environment* 77: 125–41.
Cobb, Dick, Ruth Feber, Alan Hopkins, Liz Stockdale, Tim O'Riordan, Bob Clements, Les Firbank, Keith Goulding, Steve Jarvis, and David Macdonald. 1999. Integrating the Environmental and Economic Consequences of Converting to Organic Agriculture: Evidence from a Case Study. *Land Use Policy* 16(4): 207–21.
Cochrane, Willard. 1993. *Development of American Agriculture: An Historical Analysis*. 2d ed. Minneapolis: University of Minnesota Press.
Coleman, Eliot. 2002. Beyond Organic. *Mother Earth News*, December/January, 73–74.
Colla, Giuseppe, Jeffrey P. Mitchell, Durga D. Poudel, and Steve R. Temple. 2002. Changes of Tomato Yield and Fruit Elemental Composition in Conventional, Low Input, and Organic Systems. *Journal of Sustainable Agriculture* 20(2): 53–67.
Commoner, Barry. 1992. *Making Peace with the Planet*. New York: New Press.
Conacher, Jeanette, and Arthur Conacher. 1998. Organic Farming and the Environ-

ment, with Particular Reference to Australia: A Review. *Biological Agriculture and Horticulture* 16: 145–71.

Cone, Cynthia Abbott, and Andrea Myhre. 2000. Community-Supported Agriculture: A Sustainable Alternative to Industrial Agriculture? *Human Organization* 59(2): 187–97.

Coombes, Brad, and Hugh Campbell. 1998. Dependent Reproduction of Alternative Modes of Agriculture: Organic Farming in New Zealand. *Sociologia Ruralis* 38(2): 127–45.

Cornell University. 2001. Genetically Engineered Foods: A Consumer Guide to What's in Store. http://www.geo-pie.cornell.edu (accessed October 24, 2003).

———. 2003. GE Foods in the Market. Genetically Engineered Organisms, Public Issues Education Project. http://www.geo-pie.cornell.edu. (accessed October 24, 2003).

Costanza, Robert, Ralph d'Arge, Rudolf de Groot, Stephen Farber, Monica Grasso, Bruce Hannon, Karin Limburg, Shahid Naeem, Robert O'Neill, Jose Paruelo, Robert G. Raskin, Paul Sutton, and Marjan van den Belt. 1997. The Value of the World's Ecosystem Services and Natural Capital. *Nature* 387: 253–60.

Creamer, Nancy. 2001. Myth vs. Reality: Avery's Rhetoric Meets the Real World of Organic. Organic Farming Research Foundation *Information Bulletin*, Summer, 1, 6, 7.

Cummings, Claire Hope. 2004. Silent Winter? *World Watch Magazine*, May/June, 10–17.

Cummins, Ronnie. 2001. Biotech Bullies: The Debate Intensifies. BioDemocracy News #33, Organic Consumers Association. http://www.organicconsumers.org/newsletter/biod33.cfm. (accessed January 6, 2004).

Cummins, Ronnie, and Ben Lilliston. 2000. Genetically Engineered Food: A Self-Defense Guide for Consumers. New York: Marlowe.

Curl, Cynthia L., Richard A. Fenske, and Kai Elgethun. 2003. Organophosphorus Pesticide Exposure of Urban and Suburban Preschool Children with Organic and Conventional Diets. *Environmental Health Perspectives* 111(3): 377–82.

Dalecki, Michael G., and Bob Bealer. 1984. Who Is the "Organic" Farmer? *Rural Sociology* 4(1): 11–18.

Dalgaard, Tommy, Niels Halberg, and John Porter. 2001. A Model for Fossil Energy Use in Danish Agriculture Used to Compare Organic and Conventional Farming. *Agriculture, Ecosystems and Environment* 87: 51–67.

Delate K., M. Duffy, C. Chase, A. Holste, H. Friedrich, and N. Wantate. 2003. An Economic Comparison of Organic and Conventional Grain Crops in a

Long-Term Agroecological Research (LTAR) Site in Iowa. *American Journal of Alternative Agriculture* 18(2): 59–69.
DeLind, Laura B. 1994. Organic Farming and Social Context: A Challenge for Us All. *American Journal of Alternative Agriculture* 9(4): 146–47.
———. 1998. Close Encounters with a CSA: The Reflections of a Bruised and Somewhat Wiser Anthropologist. *Agriculture and Human Values* 16: 3–9.
———. 2000. Transforming Organic Agriculture into Industrial Organic Products: Reconsidering National Organic Standards. *Human Organization* 59: 198–208.
den Biggelaar, Christoffel, and Murari Suvedi. 2000. Farmers' Definitions, Goals, and Bottlenecks of Sustainable Agriculture in the North-Central Region. *Agriculture and Human Values* 17(4): 347–58.
Dimitri, Carolyn, and Catherine Greene. 2002. Recent Growth Patterns in the U.S. Organic Foods Market. USDA, Economic Research Service, Agriculture Information Bulletin no. 777.
Dimitri, Carolyn, and Nessa J. Richman. 2000. Organic Food Markets in Transition. Henry A. Wallace Center for Agricultural and Environmental Policy. Policy Studies Report 14. Greenbelt MD: Winrock International.
Dixon, Mary Lou, Anthony Tyson, and Edward A. Brown. 1992. Your Drinking Water: Pesticides. University of Georgia College of Agricultural and Environmental Sciences Circular 819–6. Cooperative Extension Service. http://www.ces.uga.edu/pubcd/C819–6W.html (accessed January 6, 2004).
Dobbs, T. L., M. Leddy, and J. Smolik. 1988. Factors Influencing the Economic Potential for Alternative Farming Systems: Case Analyses in South Dakota. *American Journal of Alternative Agriculture* 3(1): 26–34.
Dobbs, T. L., R. C. Shane, and D. M. Feuz. 2000. Lessons Learned from the Upper Midwest Organic Marketing Project. *American Journal of Alternative Agriculture* 15(3): 119–28.
DuPuis, E. Melanie. 2000. Not in My Body: rBGH and the Rise of Organic Milk. *Agriculture and Human Values* 17: 285–95.
Duram, Leslie A. 1997. A Pragmatic Study of Conventional and Alternative Farmers in Colorado. *Professional Geographer* 49(2): 202–13.
———. 1998a. Organic Agriculture in the United States: Current Status and Future Regulation. *Choices: Food, Farm, and Resource Issues*, 2nd Quarter, 34–38.
———. 1998b. Taking a Pragmatic Behavioral Approach to Alternative Agriculture Research. *American Journal of Alternative Agriculture* 13(2): 90–97.
———. 2000. Agent's Perceptions of Structure: How Illinois Organic Farmers View Political, Economic, Social, and Ecological Factors. *Agriculture and Human Values* 17: 35–48.

Duram, Leslie A., and Kelli L. Larson. 2001. Agricultural Research and Alternative Farmers' Information Needs. *Professional Geographer* 53(1): 84–96.
Edwards-Jones, G., and O. Howells. 2001. The Origin and Hazard of Inputs to Crop Protection in Organic Farming Systems: Are They Sustainable? *Agricultural Systems* 67(1): 31–47.
Egri, Carolyn. 1999. Attitudes, Backgrounds, and Information Preferences of Canadian Organic and Conventional Farmers: Implications for Organic Farming Advocacy and Extension. *Journal of Sustainable Agriculture* 13(3): 45–72.
Entz, M. H., R. Guilford, and R. Gulden. 2001. Crop Yield and Soil Nutrient Status on Fourteen Organic Farms in the Eastern Portion of the Northern Great Plains. *Canadian Journal of Plant Science* 81(2): 351–54.
Environmental Working Group. 2004. *Point and Click Produce Aisle* and *Foods You'll Want To Buy Organic.* http://www.ewg.org/foodnews (accessed May 10, 2004).
European Union. 1999. Council Decision of 17 December 1999 Concerning BST. 1999/879/EC. http://europa.eu.int/eur-lex (access May 14, 2004).
Fairweather, John R. 1999. Understanding How Farmers Choose between Organic and Conventional Production: Results from New Zealand and Policy Implications. *Agriculture and Human Values* 16(1): 51–63.
Fass, Allison. 2002. Fieldwork. *Forbes* 170(11): 248.
Feenstra, Gail 1997. Local Food Systems and Sustainable Communities. *American Journal of Alternative Agriculture* 12(1): 28–36.
———. 2002. Creating Space for Sustainable Food Systems: Lessons from the Field. *Agriculture and Human Values* 19(2): 99–106.
Fetter, T. Robert, and Julie A. Caswell. 2002. Variation in Organic Standards Prior to the National Organic Program. *American Journal of Alternative Agriculture* 17(2): 55–74.
Fieldhouse, Paul. 1996. Community Shared Agriculture. *Agriculture and Human Values* 13(3): 43–47.
Foster, Carolyn, and Nicolas Lampkin. 1999. *European Organic Production Statistics, 1993–1996. Organic Farming in Europe: Economics and Policy*. Vol. 3. Stuttgart: Universität Hohenheim.
Foster, Gary S., and James D. Miley. 1983. Organic Farmers and Organic Nonfarmers: The Social Context of Organic Agriculture. *Rural Sociology* 3(1): 16–21.
Fotopoulos, Christos, A. Krystallis, and M. Ness. 2003. Wine Produced by Organic Grapes in Greece: Using Means-End Chains Analysis to Reveal Organic Buyers' Purchasing Motives in Comparison to the Non-Buyers. *Food Quality and Preference* 14(7): 549–66.
Francis, C., G. Lieblein, S. Gliessman, T. Breland, N. Creamer, R. Harwood, L. Salomonsson, J. Helenius, D. Rickerl, R. Salvador, M. Wiedenhoeft,

S. Simmons, P. Allen, M. Altieri, C. Flora, and R. Poincelot. 2003. Agroecology: The Ecology of Food Systems. *Journal of Sustainable Agriculture* 22(3): 99–118.

Gardyn, Rebecca. 2002. The Big O: Organic Foods and Beverages Have Gen Ys and Boomers Salivating. *American Demographics*, October, 20.

Gaskell, Mark, Benny Fouche, Steve Koike, Tom Lanini, Jeff Mitchell, and Richard Smith. 2000. Organic Vegetable Production in California-Science and Practice. *Hort Technology* 10(4): 699–713.

Gerhardt, R. A. 1997. A Comparative Analysis of the Effects of Organic and Conventional Farming Systems on Soil Structure. *Biological Agriculture and Horticulture* 14(2): 139–57.

Gleick, James. 1987. *Chaos: Making a New Science*. New York: Penguin Books.

Glickman, Dan. 2000. Release of Final National Organic Standards. Press release no. 0426.00 (as prepared for delivery by the secretary of agriculture. Washington DC, December 20, 2000. http://www.ams.usda.gov/oldnop/glickremarks.htm (accessed January 6, 2004).

Gliessman, S. R. 1998. *Agroecology: Ecological Processes in Sustainable Agriculture*. Chelsea MI: Ann Arbor Press.

Goldberg, Adam. 2002. Consumers Union Research Team Shows: Organic Foods Really DO Have Less Pesticides. May 8, 2002. http://www.consumersunion.org/food/organicpr.htm (accessed January 6, 2004).

Goldman, Barbara, and Katherine Clancy. 1991. A Survey of Organic Produce Purchases and Related Attitudes of Food Cooperative Shoppers. *American Journal of Alternative Agriculture* 6(2): 89–96.

Goodman, David. 2000. Organic and Conventional Agriculture: Materializing Discourse and Agro-Ecological Managerialism. *Agriculture and Human Values* 17: 215–19.

Greene, Catherine. 2001. U.S. Organic Farming Emerges in the 1990s: Adoption of Certified Systems. USDA, Economic Research Service, Resource Economics Division. Agriculture Information Bulletin no. 770.

Greene, Catherine, and Amy Kremen. 2003. U.S. Organic Farming in 2000–2002: Adoption of Certified Systems. USDA, Economic Research Service, Agriculture Information Bulletin no. 780.

Grey, Mark. 2000. The Industrial Food Stream and Its Alternatives in the United States: An Introduction. *Human Organization* 59(2): 143–50.

Grinder-Pedersen, Lisbeth, S. Rasmussen, S. Bügel, L. Jørgensen, L. Dragsted, V. Gundersen, and B. Sandström. 2003. Effect of Diets Based on Foods from Conventional versus Organic Production on Intake and Excretion of Flavonoids and Markers of Defense in Humans. *Journal of Agricultural and Food Chemistry* 51(19): 5671–76.

Grogan, John, and Carol Long. 2000. The Problem with Genetic Engineering. Special Report. *Organic Gardening* (Rodale Press) 41(1): 42–47.

Grossman, J. M. 2003. Exploring Farmer Knowledge of Soil Processes in Organic Coffee Systems in Chiapas, Mexico. *Geoderma* 111(3): 267–87.

Growing for Market. 2002. USDA Takes Control of "Organic." Lawrence KS. *Growing for Market* 11(11): 1, 4–5.

Guthman, Julie. 1998. Regulating Meaning, Appropriating Nature: The Codification of California Organic Agriculture. *Antipode* 30(2): 135–56.

———. 2000. Raising Organic: An Agro-ecological Assessment of Grower Practices in California. *Agriculture and Human Values* 17(3): 257–66.

Hall, Alan, and Veronika Mogyorody. 2001. Organic Farmers in Ontario: An Examination of the Conventionalization Argument. *Sociologia Ruralis* 41(4): 399–422.

Hallman, W., W. C. Hebden, H. Aquino, C. Cuite, and J. Lang. 2003. Public Perceptions of Genetically Modified Foods: A National Study of American Knowledge and Opinion. Rutgers Food Policy Institute. RR-1003–004. New Brunswick NJ: Rutgers.

Halweil, Brian. 2000. Where Have All the Farmers Gone? *World Watch* 13(5): 12–28.

———. 2001. Organic Gold Rush. *World Watch*, May/June, 22–32.

Hamilton, Pixie, and Dennis R. Helsel. 1995. Effects of Agriculture on Groundwater Quality in Five Regions of the United States. *Ground Water* 33(2): 217–26.

Hanson, James, Erik Lichtenberg, and Steven Peters. 1997. Organic versus Conventional Grain Production in the Mid-Atlantic: An Economic and Farming System Overview. *American Journal of Alternative Agriculture* 12(1): 2–9.

Hardeng, F., and V. L. Edge. 2001. Mastitis, Ketosis, and Milk Fever in Thirty-one Organic and Ninety-three Conventional Norwegian Dairy Herds. *Journal of Dairy Science* 84(12): 2673–79.

Harris, C., S. Powers, and F. Buttel. 1980. Myth and Reality in Organic Farming: A Profile of Conventional and Organic Farmers in Michigan. *Newsline* 8: 33–43.

Heaton, Shane. 2001. Organic Farming, Food Quality, and Human Health Report. Briefing Sheet. UK Soil Association. http://www.soilassociation.org/web/sa/saweb.nsf/librarytitles/Briefing_Sheets03082001a (accessed January 6, 2004).

Heffernan, William. 1999. Consolidation in the Food and Agriculture System. Research Report to the National Farmers Union. www.nfu.org (accessed January 6, 2004).

———. 2000. "Concentration of Ownership and Control in Agriculture." In *Hungry for Profit*, ed. F. Magdoff, J. B. Foster, and F. Buttel, 61–75. New York: Monthly Review Press.

Helliker, Kevin. 2002. In Natural Foods, a Big Name's No Big Help. *Wall Street Journal*, June 7, B1, B4.

Helmers, Glenn A., Michael R. Langemeier, and Joseph Atwood. 1986. An Economic Analysis of Alternative Cropping Systems for East-Central Nebraska. *American Journal of Alternative Agriculture* 1(4): 153–58.

Henderson, Elizabeth. 2000. "Rebuilding Local Food Systems from the Grassroots Up." In *Hungry for Profit*, ed. F. Magdoff, J. B. Foster, and F. Buttel, 175–88. New York: Monthly Review Press.

Henderson, Elizabeth, and Robyn Van En. 1999. *Sharing the Harvest: A Guide to Community-Supported Agriculture*. White River Junction VT: Chelsea Green.

Hendriks, K., D. J. Stobbelaar, and J. D. van Mansvelt. 2000. The Appearance of Agriculture: An Assessment of the Quality of Landscape of Both Organic and Conventional Horticultural Farms in West Friesland. *Agriculture, Ecosystems and Environment* 77: 157–75.

Hendrickson, Mary K., and William D. Heffernan. 2002a. Concentration of Agricultural Markets: February 2002 Update. Report to the National Farmers Union. www.nfu.org (accessed January 6, 2004).

———. 2002b. Opening Spaces through Relocalization: Locating Potential Resistance in the Weaknesses of the Global Food System. *Sociologia Ruralis* 42(4): 347–69.

Hendrickson, Mary, W. Heffernan, P. Howard, and J. Heffernan. 2001. Consolidation in Food Retailing and Dairy: Implications for Farmers and Consumers in a Global Food System. Research Report to the National Farmers Union. www.nfu.org.

Hermansen, John E. 2003. Organic Livestock Production Systems and Appropriate Development in Relation to Public Expectations. *Livestock Production Science* 80(1–2): 3–15.

Hightower, Jim. 2003. *Thieves in High Places: They've Stolen Our Country – and It's Time to Take It Back*. New York: Viking.

Høgh-Jensen, Henning. 1998. Systems Theory as a Scientific Approach towards Organic Farming. *Biological Agriculture and Horticulture* 16(1): 37–52.

Honisch, M., C. Hellmeier, and K. Weiss. 2002. Response of Surface and Subsurface Water Quality to Land Use Changes. *Geoderma* 105(3–4): 277–98.

Howard, Phil. 2004. Organic Industry Structure. February. The New Farm, Rodale Institute. www.newfarm.org.

Hyvönen, Terho, E. Ketoja, J. Salonen, H. Jalli, and J. Tiainen. 2003. Weed Species Diversity and Community Composition in Organic and Conventional Cropping of Spring Cereals. *Agriculture Ecosystems and Environment* 97 (1–3): 131–49.

Ikerd, John E. 1997. Toward an Economics of Sustainability. Department of Agricultural Economics, University of Missouri. http://www.ssu.missouri.edu/faculty/JIkerd/papers/econ-sus.htm (accessed January 6, 2004).

Ilbery, Brian. 1978. Agricultural Decision Making: A Behavioural Perspective. *Progress in Human Geography* 2(3): 448–66.

Ilbery, Brian, Lewis Holloway, and Ruth Arber. 1999. The Geography of Organic Farming in England and Wales in the 1990s. *Tijdschrift voor economische en Sociale Geografie* 90(3): 285–95.

Ilbery, Brian, and M. Kneafsey. 1999. Niche Markets and Regional Specialty Food Products in Europe: Towards a Research Agenda. *Environment and Planning A* 31: 2207–22.

Illinois Stewardship Alliance. 2002. Stewardship Farm: Final Report on the Farming Systems Comparison Study: Comparing the Economics of Conventional, No Till, Organic, and Three Crop Farming Systems. Rochester, Ill.

International Center for Technology Assessment. 2001. Pioneer Hi-Bred International v. J.E.M. Ag Supply, Farm Advantage, et al. 534 U.S. Analysis of Opinion. http://www.icta.org/intelprop/FarmAdAnalysis.pdf (accessed January 6, 2004).

Jackson, Dana, and Laura Jackson, eds. 2002. *The Farm as Natural Habitat: Reconnecting Food Systems with Ecosystems*. Washington DC: Island Press.

Jaffe, Gregory A. 2001. Lessen the Fear of Genetically Engineered Crops. *Christian Science Monitor*, August 8, 8.

Jawson, Michael D., and Carolee T. Bull. 2002. USDA Research into Organic Farming. *American Journal of Alternative Agriculture* 17(4): 201–2.

Jolly, Desmond A., and Kim Norris. 1991. Marketing Prospects for Organic and Pesticide-Free Produce. *American Journal of Alternative Agriculture* 6(4): 174–79.

Jolly, Desmond A., Howard G. Schutz, Katherine V. Diaz-Knauf, and Jagjeet Johal. 1989. Organic Foods: Consumer Attitudes and Use. *Food Technology* 43(11): 60–66.

Kaltoft, Pernille. 1999. Values about Nature in Organic Farming Practice and Knowledge. *Sociologia Ruralis* 39(1): 39–53.

———. 2001. Organic Farming in Late Modernity: At the Frontier of Modernity or Opposing Modernity. *Sociologia Ruralis* 41(1): 146–58.

Kaufman, Marc. 2000. U.S. Sets "Organic" Standard: USDA Seal to Debut on Foods in 2001. *Washington Post*, December 21.

Kersebaum, K. C., J. Steidl, O. Bauer, and H. Piorr. 2003. Modeling Scenarios to Assess the Effects of Different Agricultural Management and Land Use Options to Reduce Diffuse Nitrogen Pollution into the River Elbe. *Physics and Chemistry of the Earth* 28(12–13): 537–45.

Kimbrell, Andrew, ed. 2002. *The Fatal Harvest Reader: The Tragedy of Industrial Agriculture*. Washington DC: Island Press.

Klonsky, Karen. 2000. Forces Impacting the Production of Organic Foods. *Agriculture and Human Values* 17(3): 233–43.

Klonsky, Karen, and Laura Tourte. 1998. Organic Agricultural Production in the United States: Debates and Directions. *American Journal of Agricultural Economics* 80(5): 1119–24.

Kloppenburg, Jack, Jr., John Hendrickson, and George Stevenson. 1996. Coming into the Foodshed. *Agriculture and Human Values* 13: 33–42.

Kouba, Maryline. 2003. Quality of Organic Animal Products. *Livestock Production Science* 80: 33–40.

Krebs, A. V. 1992. *The Corporate Reapers: The Book of Agribusiness*. Washington DC: Essential Books.

Kremen, Claire, N. Williams, and R. Thorp. 2002. Crop Pollination from Native Bees at Risk from Agricultural Intensification. *Proceedings of the National Academy of Sciences* 99(26): 16812–16.

Kuchler, Fred, Ram Chandran, and Katherine Ralston. 1996. The Linkage between Pesticide Use and Pesticide Residues. *American Journal of Alternative Agriculture* 11(4): 161–67.

Kuiper, Juliëtte. 2000. A Checklist Approach to Evaluate the Contribution of Organic Farms to Landscape Quality. *Agriculture, Ecosystems and Environment* 77: 143–56.

Lampkin, Nicolas. 1990. *Organic Farming*. Ipswich: Farming Press.

Lampkin, Nicholas, and S. Padel, eds. 1994. *The Economics of Organic Farming: An International Perspective*. Wallingford: CAB International.

Langer, Vibike. 2002. Changes in Farm Structure Following Conversion to Organic Farming in Denmark. *American Journal of Alternative Agriculture* 17(2): 75–82.

Larson, Kelli L., and Leslie A. Duram. 2000. Information Dissemination in Alternative Agricultural Research: An Analysis of Researchers in the North Central Region. *American Journal of Alternative Agriculture* 15(4): 171–80.

La Trobe, Helen L., and Tim G. Acott. 2000. Localising the Global Food System. *International Journal of Sustainable Development and World Ecology* 7(4): 309–20.

Lauck, Jon. 2000. *American Agriculture and the Problem of Monopoly: The Political Economy of Grain Belt Farming, 1953–1980*. Lincoln: University of Nebraska Press.

Leon, Warren, and Caroline Smith DeWaal. 2002. *Is Our Food Safe? A Consumer's Guide to Protecting Your Health and the Environment*. New York: Three Rivers Press.

Letourneau, D. K., and B. Goldstein. 2001. Pest Damage and Arthropod Community Structure in Organic vs. Conventional Tomato Production in California. *Journal of Applied Ecology* 38: 557–70.

Liebig, Mark, and J. Doran. 1998. Impact of Organic Production Practices on Soil Quality Indicators for Select Farms in Nebraska and North Dakota. USDA: Agricultural Research Service. 1998–06-11. http://warp.nal.usda.gov:80/tt ic/tektran/data/000009/23/0000092349.html (accessed January 6, 2004).

Lipson, Elaine Marie. 2001. *The Organic Foods Sourcebook*. New York: Contemporary Books.

Lipson, Mark. 1997. *Searching for the "O-Word": Analyzing the USDA Current Research Information System for Pertinence to Organic Farming*. Santa Cruz CA: Organic Farming Research Foundation.

Lockeretz, William. 1991. Information Requirements of Reduced-Chemical Production Methods. *American Journal of Alternative Agriculture* 6(2): 97–103.

———. 1995. Organic Farming in Massachusetts: An Alternative Approach to Agriculture in an Urbanized State. *Journal of Soil and Water Conservation* 50(6): 663–67.

———. 1997. Diversity of Personal and Enterprise Characteristics among Organic Growers in the Northeastern United States. *Biological Agriculture and Horticulture* 14: 13–24.

Lockeretz, William, and Patrick Madden. 1987. Midwestern Organic Farming: A Ten-Year Follow-Up. *American Journal of Alternative Agriculture* 11(2): 57–63.

Lockeretz, William, Georgia Shearer, Robert Klepper, and Susan Sweeney. 1978. Field Crop Production on Organic Farms in the Midwest. *Journal of Soil and Water Conservation* 33: 130–34.

Lockeretz, William, Georgia Shearer, and Daniel H. Kohl. 1981. Organic Farming in the Corn Belt. *Science* 211: 540–47.

Lockie, Stewart. 2002. "The Invisible Mouth": Mobilizing "the Consumer" in Food Production-Consumption Networks. *Sociologia Ruralis* 42(4): 278–94.

Lockie, Stewart, Kristen Lyons, Geoffrey Lawrence, and Kerry Mummery. 2002. Eating "Green": Motivations behind Organic Food Consumption in Australia. *Sociologia Ruralis* 42(1): 23–40.

Lohr, Luanne. 1998. Implications of Organic Certification for Market Structure and Trade. *American Journal of Agricultural Economics* 80(5): 1125–29.

Lohr, Luanne, and Timothy Park. 1995. Supply Elasticities and Responses to Relative Price Changes in Organic Produce Markets. *Journal of Sustainable Agriculture* 6(1): 23–45.

———. 2002. Choice of Insect Management Portfolios by Organic Farmers: Lessons and Comparative Analysis. *Ecological Economics* 43(1): 87–99.

Lohr, Luanne, and Lennart Salomonsson. 2000. Conversion Subsidies for Organic Production: Results from Sweden and Lessons for the United States. *Agricultural Economics* 22: 133–46.

Lotter, Donald W. 2003. Organic Agriculture. *Journal of Sustainable Agriculture* 21(4): 59–128.

Lu, Yao-Chi, John R. Teasdale, and Wen-Yuen Huang. 2003. An Economic and Environmental Tradeoff Analysis of Sustainable Agriculture Cropping Systems. *Journal of Sustainable Agriculture* 22(3): 25–41.

Lundegårdh, Bengt, and Anna Mårtensson. 2003. Organically Produced Plant Foods: Evidence of Health Benefits. *Acta Agricultur Scandinavica, Section B: Soil and Plant Sciences* 53(1): 3–15.

Lynggaard, Kennet. 2001. The Farmer within an Institutional Environment: Comparing Danish and Belgian Organic Farming. *Sociologia Ruralis* 41(1): 85–111.

Lyson, T. A., G. Gillespie, and D. Hilchey. 1995. Farmers' Markets and the Local Community: Bridging the Formal and Informal Economy. *American Journal of Alternative Agriculture* 10(3): 108–13.

MacIntosh D. L., J. D. Spengler, H. Ozkaynak, L. Tsai, and P. B. Ryan. 1996. Dietary Exposures to Selected Metals and Pesticides. *Environmental Health Perspectives* 104(2): 202–9.

Mäder, Paul, Andreas Fließbach, David Dubois, Lucie Gunst, Padrout Fried, and Urs Niggli. 2002. Soil Fertility and Biodiversity in Organic Farming. *Science* 296: 1694–97.

Magdoff, F., J. B. Foster, and F. Buttel, eds. 2000. *Hungry for Profit: The Agribusiness Threat to Farmers, Food, and the Environment.* New York: Monthly Review Press.

Magnusson, Maria, K. A. Arvola, U. Koivisto Hursti, L. Åberg, and P. Sjöden. 2003. Choice of Organic Foods Is Related to Perceived Consequences for Human Health and to Environmentally Friendly Behaviour. *Appetite* 40(2): 109–17.

Marshall, Andrew. 2000. Sustaining Sustainable Agriculture: The Rise and Fall of the Fund for Rural America. *Agriculture and Human Values* 17: 267–77.

Matson, P. A., W. J. Parton, A. G. Power, and M. J. Swift. 1997. Agricultural Intensification and Ecosystem Properties. *Science* 277: 504–9.

McCann, Elizabeth, Shannon Sullivan, Donna Erickson, and Raymond De Young. 1997. Environmental Awareness, Economic Orientation, and Farming Practices: A Comparison of Organic and Conventional Farmers. *Environmental Management* 21(5): 747–58.

McConnell, Campbell R. 1984. *Economics: Principles, Problems, and Policies.* 9th ed. New York: McGraw-Hill.

Michelsen, Johannes. 1996. Organic Farmers and Conventional Distribution Systems: The Recent Expansion of the Organic Food Market in Denmark. *American Journal of Alternative Agriculture* 11(1): 18–23.

———. 2001a. Organic Farming in a Regulatory Perspective: The Danish Case. *Sociologia Ruralis* 41(1): 62–84.

———. 2001b. Recent Development and Political Acceptance of Organic Farming in Europe. *Sociologia Ruralis* 41(1): 3–20.

Milestad, Rebecka, and Ika Darnhofer. 2003. Building Farm Resilience: The Prospects and Challenges of Organic Farming. *Journal of Sustainable Agriculture* 22(3): 81–97.

Mills, Paul K., and Richard Yang. 2003. Prostate Cancer Risk in California Farm Workers. *Journal of Occupational and Environmental Medicine* 45(3): 249–58.

Misra, Sukant, Chung L. Huang, and Stephen L. Ott. 1991. Georgia Consumers' Preference for Organically Grown Fresh Produce. *Journal of Agribusiness* 9(2): 53–65.

Moore, Monica. 2002. "Hidden Dimensions of Damage: Pesticides and Health." In *Fatal Harvest: The Tragedy of Industrial Agriculture*, ed. A. Kimbrell, 130–47. Washington DC: Island Press.

Mulder, Ch., D. DeZwart, H. Van Wijnen, A. Schouten, and A. Breure. 2003. Observational and Simulated Evidence of Ecological Shifts within Soil Nematode Community of Agroecosystems under Conventional and Organic Farming. *Functional Ecology* 17: 516–25.

Myhre, Anne, and Terje Traavik. 2003. Genetically Modified (GM) Crops: Precautionary Science and Conflicts of Interest. *Journal of Agricultural and Environmental Ethics* 16: 227–47.

National Campaign for Sustainable Agriculture, Inc. 2002. Organic Integrity under Attack. Action Alert. May 30, 2002. http://www.sustainableagriculture.net/a1682.php.

———. 2003. What Is Sustainable Agriculture? http://www.sustainableagriculture.net/ (accessed January 6, 2004).

National Research Council. 1996. *Colleges of Agriculture at the Land Grant Universities: Public Service and Public Policy*. Executive Summary. Committee on the Future of Land Grant Colleges of Agriculture. Washington DC: National Academy Press. http://www.nap.edu/html/landgrant/contents.html (accessed January 6, 2004).

Natural Foods Merchandiser. 2002. News, Trends, and Ideas for the Business of Natural Products. June.

Nature. 2004. Organic Farming Enters the Mainstream. *Nature*, April 22, 783. *www.nature.com*.

Nega, Eva, Roswitha Ulrich, Sigrid Werner, and Marga Jahn. 2003. Hot Water

Treatment of Vegetable Seed: An Alternative Seed Treatment Method to Control Seed-Borne Pathogens in Organic Farming. *Zeitschrift für Pflanzenkrankheiten und Pflanzenschutz* 110(3): 220–34.

Nestle, Marion. 2002. *Food Politics: How the Food Industry Influences Nutrition and Health*. Berkeley: University of California Press.

Newton, Jon. 2002. *Profitable Organic Farming*. 2d ed. Oxford: Blackwell Science.

New York Times. 2002. After Study, Zambia Rejects Altered Food. October 30.

———. 2003. Roundup Unready. February 19.

———. 2003. Staying Organic. March 5.

Norberg-Hodge, Helena. 1995. From Catastrophe to Community. *Resurgence* 171: 12–14.

Novello, Antonia. 1991. A Charge to the Conference. *Papers and Proceedings of the Surgeon General's Conference on Agricultural Safety and Health*, 48–54. Des Moines IA, April 30.

OECD. 2002. Organisation for Economic Co-operation and Development. *Workshop on Organic Agriculture*. Presentation by Wilfrid Legg.

Organic Europe. 2003. Research Institute of Organic Agriculture, Frick, Switzerland. European Commission, Agriculture Directorate-General. http://www.organic-europe.net/ (accessed January 6, 2004).

O'Riordan, Tim, and Dick Cobb. 2001. Assessing the Consequences of Converting to Organic Agriculture. *Journal of Agriculture Economics* 52(1): 22–35.

Orris, Peter, Lin Kaatz Chary, Karen Perry, and Joe Asbury. 2000. Persistent Organic Pollutants (POPS) and Human Health. World Federation of Public Health Associations' Persistent Organic Pollutants Project. Washington DC. http://www.apha.org/wfpha/popsfinal1.pdf (accessed January 6, 2004).

Pacini, Cesare, A. Wossink, G. Giesen, C. Vazzana, and R. Huirne. 2003. Evaluation of Sustainability of Organic, Integrated, and Conventional Farming Systems: A Farm and Field-Scale Analysis. *Agriculture, Ecosystems and Environment* 95(1): 273–88.

Padel, Susanne. 2001. Conversion to Organic Farming: A Typical Example of the Diffusion of an Innovation? *Sociologia Ruralis* 41(1): 40–61.

Padel, Suzanne, Nic Lampkin, and Carolyn Foster. 1999. Influence of Policy Support on the Development of Organic Farming in the European Union. *International Planning Studies* 4(3): 303–15.

Pallant, Eric, David Lansky, Jessica Rio, Lawrence Jacobs, George Schuler, and Walter Whimpenny. 1997. Growth of Corn Roots under Low-Input and Conventional Farming Systems. *American Journal of Alternative Agriculture* 12(4): 173–77.

PANNA. 2003. Pesticide Action Network of North America. Projects and Campaigns: Genetic Engineering. http://www.panna.org/campaigns/ge.html (accessed January 6, 2004).

Paulsen, Hans Marten, Uwe Volkgenannt, and Ewald Schnug. 2002. Contribution of Organic Farming to Marine Environmental Protection. *Landbauforschung Volkenrode* 52(4): 211–18.

Petrini, Carlos (with Ben Watson), eds. 2001. *Slow Food: Collected Thoughts on Taste, Tradition, and the Honest Pleasures of Food.* White River Junction VT: Chelsea Green.

Phillipson, M. 2001. Agricultural Law: Containing the GM Revolution. *Biotechnology and Development Monitor* 48: 2–5.

Pickrell, John. 2002. Federal Government Launches Organic Standards. *Science News On-line*, November 1. http://www.sciencenews.org/20021102/food.asp (accessed January 6, 2004).

Pimentel, David. 1991. "The Dimensions of the Pesticide Question." In *Ecology, Economics, Ethics: The Broken Circle*, ed. F. Herbert Bormann and Stephen R. Kellert, 59–69. New Haven: Yale University Press.

———. 1993. Economics and Energetics of Organic and Conventional Farming. *Journal of Agricultural and Environmental Ethics* 6: 53–60.

Pimentel, David, and Hugh Lehman, eds. 1993. *The Pesticide Question: Environment, Economics, and Ethics.* New York: Chapman and Hall.

Pimentel, David, M. S. Hunter, J. A. LaGro, R. A. Efroymson, J. C. Landers, F. T. Mervis, C. A. McCarthy, and A. E. Boyd. 1989a. Benefits and Risks of Genetic Engineering in Agriculture. *BioScience* 39(9): 606–14.

Pimentel, David, Thomas W. Culliney, Imo W. Buttler, Douglas J. Reinemann, and Kenneth B. Beckman. 1989b. Low-Input Sustainable Agriculture Using Ecological Management Practices. *Agriculture, Ecosystems and Environment* 27: 3–24.

Pirog, Rich, Timothy Van Pelt, Kamyar Enshayan, and Ellen Cook. 2001. Food, Fuel, and Freeways: An Iowa Perspective on How Far Food Travels, Fuel Usage, and Greenhouse Gas Emissions. Report from the Leopold Center for Sustainable Agriculture, Iowa State University. www.leopold.iastate.edu/.

Pollack, Andrew. 2003. Widely Used Crop Herbicide Is Losing Weed Resistance. *New York Times*, January 14.

Pollan, Michael. 2001a. *The Botany of Desire: A Plant's-Eye View of the World.* New York: Random House.

———. 2001b. How Organic Became a Marketing Niche and a Multibillion-Dollar Industry Naturally. *New York Times Magazine*, May 13.

———. 2002. When a Crop Becomes King. Editorial. *New York Times*, July 19.

Pretty, J. N., C. Brett, D. Gee, R. Hine, C. Mason, J. Morison, H. Raven, M. Rayment, and G. van der Bijl. 2000. An Assessment of the Total External Costs of UK Agriculture. *Agricultural Systems* 65: 113–36.

Pugliese, Patrizia. 2001. Organic Farming and Sustainable Rural Development: A Multifaceted and Promising Convergence. *Sociologia Ruralis* 41(1): 112–30.

Pulleman, M., A. Jongmans, J. Marinissen, and J. Bouma. 2003. Effects of Organic versus Conventional Arable Farming on Soil Structure and Organic Matter Dynamics in a Marine Loam in the Netherlands. *Soil Use and Management* 19(20): 157–65.

Rasul, Golam, and Gopal B. Thapa. 2003. Sustainability Analysis of Ecological and Conventional Agricultural Systems in Bangladesh. *World Development* 31(10): 1721–41.

Raynolds, Laura. 2000. Re-embedding Global Agriculture: The International Organic and Fair Trade Movements. *Agriculture and Human Values* 17: 297–309.

Reed, Matthew. 2001. Fight the Future! How the Contemporary Campaigns of the UK Organic Movement Have Arisen from Their Composting of the Past. *Sociologia Ruralis* 41(1): 131–45.

Reganold, John P., Jerry D. Glover, Preston K. Andrews, and Herbert R. Hinman. 2001. Sustainability of Three Apple Production Systems. *Nature* 410: 926–30.

Reigart, J. Routt, and James R. Roberts. 1999. *Recognition and Management of Pesticide Poisonings*. 5th ed. Office of Pesticide Programs, U.S. Environmental Protection Agency. *http://www.epa.gov/oppfead1/safety/healthcare/handbook/handbook.htm.*

Rein, Bradley. 1992. Health Hazards in Agriculture – An Emerging Issue. Document 000102014. Farm Safety Fact Sheet. U.S. Department of Agriculture Extension Service. http://www.cdc.gov/nasd/docs/d001001-d001100/d001050/da001050.pdf (accessed January 6, 2004).

Reuters. 2002. USDA: Organic Foods May Be More Contamination-Prone. Remarks from Under Secretary for Food Safety, Elsa Murano, October 24.

Rickerl, D. H., and J. D. Smolik. 1990. Farming Systems' Influences on Soil Properties and Crop Yields. *Journal of Soil and Water Conservation* 45: 121–25.

Rickson, Roy E., Paul Saffigna, and Richard Sanders. 1999. Farm Work Satisfaction and Acceptance of Sustainability Goals by Australian Organic and Conventional Farmers. *Rural Sociology* 64(2): 266–83.

Rigby, Dan, and D. Cáceres. 2001. Organic Farming and the Sustainability of Agricultural Systems. *Agricultural Systems* 68: 21–40.

Rigby, Dan, Trevor Young, and Michael Burton. 2001. The Development of and Prospects for Organic Farming in the UK. *Food Policy* 26(6): 599–613.

Robinson, Ramona, Chery Smith, Helene Murray, and Jim Ennis. 2002. Promotion of Sustainably Produced Foods: Customer Response in Minnesota Grocery Stores. *American Journal of Alternative Agriculture* 17(2): 96–104.

Rosset, Peter M., and Miguel A. Altieri. 1997. Agroecology versus Input Substitution: A Fundamental Contradiction of Sustainable Agriculture. *Society and Natural Resources* 10(3): 283–95.

Rossi, Roberto, and Dionisio Nota. 2000. Nature and Landscape Production Potentials of Organic Types of Agriculture: A Check of Evaluation Criteria and Parameters in Two Tuscan Farm-Landscapes. *Agriculture, Ecosystems and Environment* 77: 53–64.

Runyan, Jack L. 1993. A Review of Farm Accident Data Sources and Research: Review of Recently Published and Current Research. Agriculture and Rural Economy Division, Economic Research Service, U.S. Department of Agriculture. Extracted from *Bibliographies and Literature of Agriculture*, no. 125. http://www.cdc.gov/nasd/docs/d001001-d001100/d001045/d001045.pdf (accessed January 6, 2004).

Rural Advancement Foundation International. 2003. Broken Promise? Monsanto Promotes Terminator Seed Technology. Action Group on Erosion, Technology, and Concentration. www.etcgroup.org (accessed October 3, 2003).

Rutenberg, Jim, and Felicity Barringer. August 13, 2000. Media: Apology Highlights ABC Reporter's Contrarian Image. *New York Times*, August 13.

Saftlas, Audrey F., Aaron Blair, Kenneth P. Cantor, Larry Hanrahan, and Henry A. Anderson. 1987. Cancer and Other Causes of Death among Wisconsin Farmers. *American Journal of Industrial Medicine* 11(2): 119–29.

Schäfer, Martina. 2003. Kundenvielfalt Erfordert Markvielfalt: eine Untersuchung der Potenziale von vier erschiedenen Bio-Einkaufsformen. *Berichte Über Landwirtschaft* 81(1): 103–27.

Schjønning, Per, Susanne Elmholt, Lars J. Munkholm, and Kasia Debosz. 2002. Soil Quality Aspects of Humid Sandy Loams as Influenced by Organic and Conventional Long-Term Management. *Agriculture, Ecosystems and Environment* 88(3): 195–214.

Schlosser, Eric. 2002. *Fast Food Nation: The Dark Side of the All-American Meal.* New York: HarperCollins.

Schneeberger, Walter, Ika Darnhofer, and Michael Eder. 2002. Barriers to the Adoption of Organic Farming by Cash-Crop Producers in Austria. *American Journal of Alternative Agriculture* 17(1): 24–31.

Schnug, Ewald, and Silvia Haneklaus. 2002. Landwirtschaftliche Produktionstechnik und Infiltration von Böeden – Beitrag des Öekologischen Landbaus zum Vorbeugenden Hochwasserschutz. *Landbauforschung Voelkenrode* 52(4): 197–203.

Schreinemachers, D. M. 2003. Birth Malformations and Other Adverse Perinatal Outcomes in Four U.S. Wheat-Producing States. *Environmental Health Perspectives* 111(9): 1259–64.

Scofield, A. M. 1986. Organic Farming – the Origin of a Name. Editorial. *Biological Agriculture and Horticulture* 4: 1–5.

Shepherd, M. A., R. Harrison, and J. Webb. 2002. Managing Soil Organic Matter – Implications for Soil Structure on Organic Farms. *Soil Use and Management* 18(3): 284–92.

Shutler, Dave, Adele Mullie, and Robert G. Clark. 2000. Bird Communities of Prairie Uplands and Wetlands in Relation to Farming Practices in Saskatchewan. *Conservation Biology* 14(5): 1441–51.

Simcox N. J., R. A. Fenske, S. A. Wolz, I. C. Lee, and D. A. Kalman. 1995. Pesticides in Household Dust and Soil: Exposure Pathways for Children of Agricultural Families. *Environmental Health Perspectives* 103(12): 1126–34.

Sligh, Michael, and Carolyn Christman. 2003. Who Owns Organic? The Global Status, Prospects, and Challenges of a Changing Organic Market. Rural Advancement Foundation International–USA. *www.rafiusa.org.*

Smith, Katherine R. 1995. Making Alternative Agriculture Research Policy? *American Journal of Alternative Agriculture* 10(1): 10–18.

Smith, Miranda, and Elizabeth Henderson, eds. 1998. *The Real Dirt: Farmers Tell about Organic and Low In-Put Practices in the Northeast.* 2d ed. Burlington VT: Northeast Region Sustainable Agriculture Research and Education Program.

Smolik, James, and Thomas Dobbs. 1991. Crop Yields and Economic Returns Accompanying the Transition to Alternative Farming Systems. *Journal of Production Agriculture* 4(2): 153–61.

Sok, Emy, and Lewrene Glaser. 2001. Tracking Wholesale Prices for Organic Produce. USDA: ERS, *Agricultural Outlook*, October.

Sooby, Jane. 2003. *State of the States: Organic Farming Systems Research at Land Grant Institutions, 2000–2001.* 2d ed. Santa Cruz CA: Organic Farming Research Foundation.

Soule, Judith D., and Jon K. Piper. 1992. *Farming in Nature's Image: An Ecological Approach to Agriculture.* Covelo CA: Island Press.

Standard and Poor's. 2003. Dean Foods. Stock Reports. 12 April 2003. *http://www.buyside.com.*

Stanhill, G. 1990. The Comparative Productivity of Organic Agriculture. *Agriculture, Ecosystems and Environment* 30: 1–26.

Steingraber, Sandra. 1997. *Living Downstream: An Ecologist Looks at Cancer and the Environment.* New York: Addison-Wesley.

Stobbelaar, Derk Jan, and Jan Diek van Mansvelt. 2000. The Process of Landscape Evaluation: Introduction to the 2d Special AGEE Issue of the Concerted Action: "The Landscape and Nature Production Capacity of Organic/Sustainable Types of Agriculture." *Agriculture, Ecosystems and Environment* 77: 1–15.

Stobbelaar, Derk Jan, Juliëtte Kuiper, Jan Diek van Mansvelt, and Emmanuiol Kabourakis. 2000. Landscape Quality on Organic Farms in the Messara Valley, Crete Organic Farms as Components in the Landscape. *Agriculture, Ecosystems and Environment* 77: 79–93.

Stolton, Sue, and B. Geier. 2002. The Relationship between Biodiversity and Organic Agriculture. Discussion document. Pan-European Conference on Agriculture and Biodiversity, Paris, June 5–7. http://www.nature.coe.int/CONF_AGRI_2002/agri03e.01.doc (accessed December 17, 2003).

Stoner, Kimberly, ed. 1998. *Alternatives to Insecticides for Managing Vegetable Insects.* Proceedings of a Farmer/Scientist Conference. Ithaca NY: Natural Resource, Agriculture, and Engineering Service, Cooperative Extension.

Storstad, Oddveig, and Hilde Bjørkhaug. 2003. Foundations of Production and Consumption of Organic Food in Norway: Common Attitudes among Farmers and Consumers? *Agriculture and Human Values* 20(2): 151–63.

Strange, Marty. 1988. *Family Farming: A New Economic Vision.* Lincoln: University of Nebraska Press.

Supermarket Strategic Alert. 2003. Special Report. Pollack Associates. http://www.supermarketalert.com/pdf%20docs/3BrandsPriLabel.pdf (accessed May 18, 2004).

Swezey, Sean, Matthew Werner, Marc Buchanan, and Jan Allison. 1998. Comparison of Conventional and Organic Apple Production Systems during Three Years of Conversion to Organic Management in Coastal California. *American Journal of Alternative Agriculture* 13(4): 162–80.

Tamm, Lucius. 2001. Organic Agriculture: Development and the State of the Art. *Journal of Environmental Monitoring* 3: 92–96.

Tate, William B. 1994. The Development of the Organic Industry and Market: An International Perspective. In *The Economics of Organic Farming*, ed. N. Lampkin and S. Padel, 11–25. Wallingford: Cab International.

Thompson, Gary D. 1998. Consumer Demand for Organic Foods: What We Know and What We Need to Know. *American Journal of Agricultural Economics* 80(5): 1113–18.

———. 2000. International Consumer Demand for Organic Foods. *Hort Technology* 10(4): 663–74.

Torjusen, Hanne, Geir Lieblein, Margareta Wandel, and Charles A. Francis. 2001. Food System Orientation and Quality Perception among Consumers and

Producers of Organic Food in Hedmark County, Norway. *Food Quality and Preference* 12(3): 207–16.

Tovey, Hilary. 1997. Food, Environmentalism, and Rural Sociology: On the Organic Movement in Ireland. *Sociologia Ruralis* 37(1): 21–37.

Trautmann, Nancy M., Keith S. Porter, and Robert J. Wagenet. 1998. Pesticides: Health Effects in Drinking Water. Natural Resources Report. Pesticide Management Education Program. Cornell Cooperative Extension. http://pmep.cce.cornell.edu/facts-slides-self/facts/pes-heef-grw85.html (accessed January 6, 2004).

Tzouvelekas, Vangelis, Christos J. Pantzios, and Christos Fotopoulos. 2001. Technical Efficiency of Alternative Farming Systems: The Case of Greek Organic and Conventional Olive-Growing Farms. *Food Policy* 26(6): 549–69.

Udall, Stewart L. 1963. *The Quiet Crisis.* New York: Avon Books.

UN Report. 2003. United Nations Panel Calls for Controls on Asbestos, Pesticides, and Lead Additives. William J. Stibravy, ICC Representative to the UN. http://www.uscib.org/index.asp?documentID=2498 (accessed January 6, 2004).

UNEP. 2003. United Nations Environmental Programme. 2003. What Are POPS? Global Program of Action for the Protection of the Marine Environment from Land-Based Activities. http://pops.gpa.unep.org/01what.htm (accessed January 6, 2004).

USDA. 2002. *Farm Bill 2002.* May 13, 2002. www.usda.gov/farmbill/ (accessed January 6, 2004).

USDA–*Amber Waves.* 2003. Information Sways Consumer Attitudes toward Biotech Foods. *Amber Waves* 1(3): 6.

USDA–AMS. 2002. Agricultural Marketing Service. *Farmers Market Facts!* http://www.ams.usda.gov/farmersmarkets/facts.htm (accessed January 6, 2004).

USDA–CSREES. 2004. Cooperative State Research, Education, and Extension Service Integrated Organic Program. *Funding Opportunities.* http://www.csrees.usda.gov/fo/fundview.cfm?fonum=1141.

USDA–ERS–GMO. 2003. Economic Research Service–Genetically Modified Organisms. *Agricultural Biotechnology: Adoption of Biotechnology and Its Production Impacts.* http://www.ers.usda.gov/Briefing/Biotechnology/chapter1.htm (accessed January 6, 2004).

USDA–FSA. 1998. Farm Service Agency. *Nonrecourse Marketing Assistance Loans and Loan Deficiency Payments.* www.fsa.usda.gov/pas/publications/facts/nonrec98.pdf (accessed January 6, 2004).

USDA–NASS. 1997. National Agricultural Statistics Service. *Census of Agriculture.* http://www.nass.usda.gov/census/ (accessed January 6, 2004).

———. 2002. *Census of Agriculture.* Form no. 02-A0201.

———. 2003. *Agricultural Statistics.* http://www.usda.gov/nass/pubs/agstats.htm (accessed January 6, 2004).

USDA–NASS–GMO. 2003. USDA–National Agricultural Statistics Service–Genetically Modified Organisms. *Acreage.* Report released June 30, 2003. http://usda.mannlib.cornell.edu/reports/nassr/field/pcp-bba/acrg0603.pdf (accessed January 6, 2004).

USDA News Release. 2002. *Veneman Marks Implementation of USDA National Organic Standards.* October 21. http://www.usda.gov/news/releases/2002/10/0453.htm (accessed January 6, 2004).

USDA–SARE. 1998. Sustainable Agriculture Research and Education Program. *Ten Years of SARE: A Decade of Programs, Partnerships, and Progress in Sustainable Agriculture Research and Education.* http://www.sare.org/htdocs/pubs/98Highlights/#Ten (accessed January 6, 2004).

———. 2000. *Naturalize Your Farming System: A Whole-Farm Approach to Managing Pests. Sustainable Agriculture Network (SAN) Information Bulletin.* http://www.sare.org/farmpest/index.htm (accessed January 6, 2004).

———. 2003. *Exploring Sustainability in Agriculture.* http://www.sare.org/bulletin/explore/ (accessed January 6, 2004).

USEPA. 1999. *Assessing Health Risks from Pesticides Fact Sheet.* http://www.epa.gov/pesticides/factsheets/riskassess.htm (accessed January 6, 2004).

USEPA–Canceled Uses. 2003. Restricted and Canceled Uses. *Pesticides: Regulating Pesticides.* http://www.epa.gov/pesticides/regulating/restricted.htm (accessed January 6, 2004).

USEPA–Compliance. 2000. Providing Safe Drinking Water in America. National Public Water Systems Compliance Report. Office of Enforcement and Compliance Assurance. EPA-305-R-02–001. http://www.epa.gov/safewater/annual/sdwcom2002.pdf (accessed January 6, 2004).

USEPA–Contaminants. 2003. *List of Contaminants and Their MCLS: National Drinking Water Regulations.* EPA 816-F-02–013. http://www.epa.gov/safewater/mcl.html#mcls (accessed January 6, 2004).

USEPA–Factoids. 2003. *Drinking Water and Ground Water Statistics for 2002.* Office of Ground Water and Drinking Water (4606M) EPA 816-K-03–001. http://www.epa.gov/safewater/data/pdfs/02factoids.pdf (accessed January 6, 2004).

USEPA–Inert Ingredients. 2003. *Inert Ingredients in Pesticide Products.* http://www.epa.gov/opprd001/inerts/ (accessed January 6, 2004).

USEPA–Standards. 2002. *Drinking Water Standards and Health Advisories.* EPA 822-R-02–038. http://www.epa.gov/ost/drinking/standards/dwstandards.pdf (accessed January 6, 2004).

USEPA–UN PIC List. 2003 United Nations Prior Informed Consent List. *List of Pesti-*

cides Banned and Severely Restricted in the United States. http://www.epa.gov/oppfead1/international/piclist.htm (accessed January 6, 2004).

USFDA. 2001. *Guidance for Industry: Voluntary Labeling Indicating Whether Foods Have or Have Not Been Developed Using Bioengineering*. United States Food and Drug Administration, Center for Food Safety and Applied Nutrition. http://www.cfsan.fda.gov/ dms/biolabgu.html (accessed January 6, 2004).

——— 2002. *Pesticide Program: Residue Monitoring 2000*. United States Food and Drug Administration, Center for Food Safety and Applied Nutrition. http://www.cfsan.fda.gov/ dms/pes00rep.html#summ4 (accessed January 6, 2004).

USFDA–GMO. 2002. *FDA List of Completed Consultations on Bioengineered Foods*. United States Food and Drug Administration, Center for Food Safety and Applied Nutrition. http://www.cfsan.fda.gov/ lrd/biocon.html (accessed January 6, 2004).

USGS. 1999. *The Quality of Our Nation's Waters – Nutrients and Pesticides*. Circular 1225. http://water.usgs.gov/pubs/circ/circ1225/pdf/front.pdf (accessed January 6, 2004).

———. 2001. *National Water-Quality Assessment (NAWQA): Informing Water-Resource Management and Protection Decisions*. http://water.usgs.gov/nawqa/docs/xrel/ (accessed January 6, 2004).

van Bruggen, Ariena H. C., and Aad J. Termorshuizen. 2003. Integrated Approaches to Root Disease Management in Organic Farming Systems. *Australasian Plant Pathology* 32(2): 141–56.

van Elsen, Thomas. 2000. Species Diversity as a Task for Organic Agriculture in Europe. *Agriculture, Ecosystems and Environment* 77: 101–9.

van Mansvelt, J. D., D. J. Stobbelaar, and K. Hendriks. 1998. Comparison of Landscape Features in Organic and Conventional Farming Systems. *Landscape and Urban Planning* 41: 209–27.

Vazquez, R. I., B. Stinner, and D. McCartney. 2003. Corn and Weed Residue Decomposition in Northeast Ohio Organic and Conventional Dairy Farms. *Agriculture, Ecosystems and Environment* 95(2–3): 559–65.

Verhoog, Henk, Mirjam Matze, Edith Lammerts van Bueren, and Ton Baars. 2003. The Role of the Concept of the Natural (Naturalness) in Organic Farming. *Journal of Agricultural and Environmental Ethics* 16(1): 29–49.

Vos, Timothy. 2000. Visions of the Middle Landscape: Organic Farming and the Politics of Nature. *Agriculture and Human Values* 17: 245–56.

Walters, Charles, and C. J. Fenzau. 1996. *Eco-Farm: An Acres U.S.A. Primer*. Austin TX: Acres U.S.A.

Walz, Erica. 1999. *Final Results of the Third Biennial National Organic Farmers' Survey*. Santa Cruz CA: Organic Farming Research Foundation.

Watson, C. A., D. Atkinson, P. Gosling, L. R. Jackson, and F. W. Rayns. 2002. Managing Soil Fertility in Organic Farming Systems. *Soil Use and Management* 18(3): 239–47.

Weir, David, and Mark Schapiro. 1981. *Circle of Poison: Pesticides and People in a Hungry World*. Oakland CA: Institute for Food and Development Policy.

Williams-Derry, Clark, and Ken Cook. 2000. *Green Acres: How Taxpayers Are Subsidizing the Demise of the Family Farm*. Environmental Working Group. http://www.ewg.org/reports/greenacres/pr.html (accessed January 6, 2004).

Woese, Katrin, Dirk Lange, Christian Boess, and Klaus Werner Bögl. 1997. A Comparison of Organically and Conventionally Grown Foods – Results of a Review of the Relevant Literature. *Journal of the Science of Food and Agriculture* 74: 281–93.

Worster, Donald. 1984. "Good Farming and the Public Good." In *Meeting the Expectations of the Land: Essays in Sustainable Agriculture and Stewardship*, ed. W. Jackson, W. Berry, and B. Colman, 31–41. San Francisco: North Point Press.

Worthington, Virginia. 1998. Effect of Agricultural Methods on Nutritional Quality: A Comparison of Organic with Conventional Crops. *Alternative Therapies* 4(1): 58–69.

———. 2001. Nutritional Quality of Organic versus Conventional Fruits, Vegetables, and Grains. *Journal of Alternative and Complimentary Medicine* 7(2): 161–73.

Xie, B., X. Wang, Z. Ding, and Y. Yang. 2003. Critical Impact Assessment of Organic Agriculture. *Journal of Agricultural and Environmental Ethics* 16(3): 297–311.

Youngberg, Garth, Neill Schaller, and Kathleen Merrigan. 1993. "The Sustainable Agriculture Policy Agenda in the United States: Politics and Prospects." In *Food for the Future*, ed. Patricia Allen, 295–318. New York: Wiley.

Zinati, Gladis M. 2002. Transition from Conventional to Organic Farming Systems: I. Challenges, Recommendations, and Guidelines for Pest Management. *Hort Technology* 12(4): 606–10.

Index

IN THE OUR SUSTAINABLE FUTURE SERIES

1
Ogallala: Water for a Dry Land
John Opie

2
Building Soils for Better Crops: Organic Matter Management
Fred Magdoff

3
Agricultural Research Alternatives
William Lockeretz and Molly D. Anderson

4
Crop Improvement for Sustainable Agriculture
Edited by M. Brett Callaway and Charles A. Francis

5
Future Harvest: Pesticide-Free Farming
Jim Bender

6
A Conspiracy of Optimism: Management of the National Forests since World War Two
Paul W. Hirt

7
Green Plans: Greenprint for Sustainability
Huey D. Johnson

8
Making Nature, Shaping Culture: Plant Biodiversity in Global Context
Lawrence Busch, William B. Lacy, Jeffrey Burkhardt, Douglas Hemken, Jubel Moraga-Rojel, Timothy Koponen, and José de Souza Silva

9
Economic Thresholds for Integrated Pest Management
Edited by Leon G. Higley and Larry P. Pedigo

10
Ecology and Economics of the Great Plains
Daniel S. Licht

11
Uphill against Water: The Great Dakota Water War
Peter Carrels

12
Changing the Way America Farms: Knowledge and Community in the Sustainable Agriculture Movement
Neva Hassanein

13
Ogallala: Water for a Dry Land, second edition
John Opie

14
Willard Cochrane and the American Family Farm
Richard A. Levins

15
Raising a Stink: The Struggle over Factory Hog Farms in Nebraska
Carolyn Johnsen

16
The Curse of American Agricultural Abundance: A Sustainable Solution
Willard W. Cochrane

17
Good Growing: Why Organic Farming Works
Leslie A. Duram